Two Manuals Combined into One Book

Military Training for Urban Operations

And

FM 7-85 Ranger Unit Operations

Department of Defense

Pentagon
Publishing

Training Circular
No. 90-1

Headquarters
Department of the Army
Washington, DC, 19 May 2008

Training for Urban Operations

Contents

Distribution Restriction: Approved for public release; distribution is unlimited.

*This publication supersedes TC 90-1, 1 April 2002.

Figures

Tables

Chapter 1
Introduction

This chapter discusses definitions, training strategy, prerequisite training, individual task training, and collective task training for urban operations (UO). Military operations on urbanized terrain (MOUT) and UO differ in ways that reflect the changes in the way we engage in war today.

Section I. DEFINITIONS

This section defines and compares MOUT and UO.

Note: The terms "urban operations" and "UO" replaces all instances of "military operations on urbanized terrain" and "MOUT."

MILITARY OPERATIONS ON URBANIZED TERRAIN

1-1. In the past, Army doctrine characterized urban areas as a type of specialized terrain, like jungles and mountains, and called operations in this type of terrain *military operations on urbanized terrain* (MOUT). Doctrine emphasized tactical warfighting tasks on this terrain, which was characterized by densely situated buildings and streets, and assumed the enemy consisted of conventional (Soviet-type) forces. Defeating or destroying those forces overrode all other considerations.

URBAN OPERATIONS

1-2. The concept of UO assumes that the urbanized terrain is populated, and that the populace must be a foremost consideration. UO recognizes that the populace can help either side, and that it requires a working infrastructure. Urban operations, whether against an insurgent or unconventional enemy, thus vastly differ from MOUT (Table 1-1).

Table 1-1. Comparison of MOUT and UO.

	MOUT	*UO*
Enemy force	Conventional, Soviet type	Insurgent or unconventional
Mission	Defeat or destroy enemy above all other considerations	Defeat the enemy Gain and retain the good will of the populace
Populace	Not considered	Important consideration

1-3. Although certain techniques and tactics are common to both types of operations, UO goes beyond MOUT, focusing not only on defeating the enemy, but also on gaining and retaining the good will of the population. UO consider the complex and dynamic interactions and relationships between the urban area's key components--the terrain (natural and manmade), the population, and the supporting infrastructure--as an overlapping and interdependent system of systems.

1-4. A single urban operation may include the full range of Army operations--offensive, defensive, stability and support (Appendix C)--that may be executed, either sequentially or simultaneously (FM 3-06, Chapter 1). Furthermore, to accomplish an objective, UO demand working with sister services, US government agencies, nongovernmental organizations, and regional and international organizations.

Section II. TRAINING STRATEGY

The UO training strategy is primarily organized by environment and phase. Training environments include live, virtual, and constructive (Figure 1-1), and are based on the unit's METL. Training phases are initial, intermediate, and culminating. Table 1-2 shows the relationship between the training environments and the first three training phases. (See also Appendix D, which discusses leader training.)

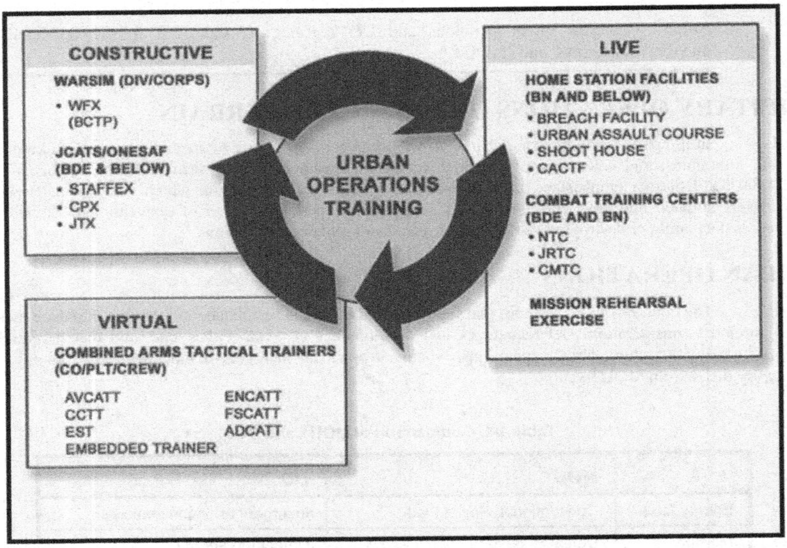

Figure 1-1. Urban Operations Training Support Strategy.

TRAINING PHASES

1-5. The three training phases are initial, intermediate, and culminating. Each of these includes some live, virtual, and constructive training.

INITIAL PHASE

1-6. This phase, which trains individual and collective tasks, is the first step in the building-block approach. It forms the basis of unit or collective training. For Soldiers and leaders, mastery of individual and crew skills is a prerequisite for UO training. The tasks train basic marksmanship skills and other Soldier tasks needed to perform any tactical operation. Included are training in a live environment (precision marksmanship) and virtual environment (Engagement Skills Trainer [EST], crew simulators, and CCTTs).

Table 1-2. Training environment/training phase matrix.

TRAINING PHASE / TRAINING ENVIRONMENT	INITIAL	INTERMED	CULMINATING
LIVE	Precision Marksmanship (Indiv / Sqd)	Breach Facility, Urban Assault Course, Shoot House (Team / Squad / Platoon)	CACTF, Platoon STX, Company STX, Battalion FTX Deployment CTC Rotation Mission Rehearsal Exercise
VIRTUAL	EST (Indiv / Sqd) Crew Simulators and CCTT (Crews Plt Ldrs, Co Cdrs)		
CONSTRUCTIVE			WARSIM (division/corps), which includes warfighting exercise (battle command training program). JCATS/OneSAF (brigade and below), staff exercises (STAFFEXs), command post exercises (CPXs), and joint training exercises (JTX)

Intermediate Phase

1-7. This phase of training is characterized by collective task training in varied conditions such as limited visibility; different entry techniques and points; incorporation of combatants, noncombatants, and OPFOR; use of MILES; and live-fire exercises. It is the vital link between individual tasks and mission accomplishment. It should be conducted at, but is not limited to, the *Breach Facility*, *Live-Fire Shoot House*, and *UAC*. (Section III discusses these facilities in detail.) Many types of organizations can use these facilities. Trainers *must* be certified before conducting demolition training IAW local standing operating procedure (SOP) or ACOM policy. AR 385-63 prescribes safety requirements. Except for Aviation, Armor, and other branches that have no practical use for these facilities, this phase must be completed to advance to the culminating phase.

Culminating Phase

1-8. The Combined Arms Collective Training Facility (CACTF, Figure 1-2) is where the culminating phase of urban operations training ends. This phase can include multiechelon, combined arms, and branch specific STX training may be conducted. Urban operations leader training occurs in leader professional development courses (Appendix D). Units have developed internal urban warfare training programs and TRADOC schools have assembled technical mobile training teams (MTTs) for marksmanship, combatives, and breaching techniques.

Education Phase

- Urban warfare reading/video list.
- Urban operations equipment.
- Review of Infantry doctrinal publications (FM 3-21.10, FM 3-06.11, for Infantry), and corresponding publications for other branches.
- Small-unit urban movement techniques and considerations.

Planning and Execution Phase

- Prepare company OPORD or WFF-related products.
- Brief platoon/company OPORD/WFF-related products to commander or staff representative.
- Participate as briefer in urban environment TEWT.

Sustainment Phase

- Participate in company/platoon TEWT or other urban operations event.
- Execute/facilitate a company or platoon TEWT.

TRAINING ENVIRONMENTS

1-9. Both live and virtual training environments are used to train UO.

Live Environment

1-10. The live training environment consists of home station training using the Urban Assault Course (UAC), Shoot House (SH), Breach Facility (BF), Combined Arms Collective Training Facility (CACTF), Combat Training Center (CTC) rotations, and any other live training that supports the UO training.

VIRTUAL ENVIRONMENT

1-11. The virtual training environment consists of simulators such as Close Combat Tactical Trainer (CCTT) or Engagement Skills Trainer (EST), which are used to train small-unit leaders and crews. The constructive environment consists of battle staff training for battalion and above staff and leaders such as Joint Conflict and Tactical Simulation (JCATS)/OneSAF.

CONSTRUCTIVE ENVIRONMENT

1-12. The constructive training environment consists of the following, which all occur in the culminating training phase:

- WARSIM (division/corps), which includes warfighting exercise (battle command training program).
- JCATS/OneSAF (brigade and below), staff exercises (STAFFEXs), command post exercises (CPXs), and joint training exercises (JTX)

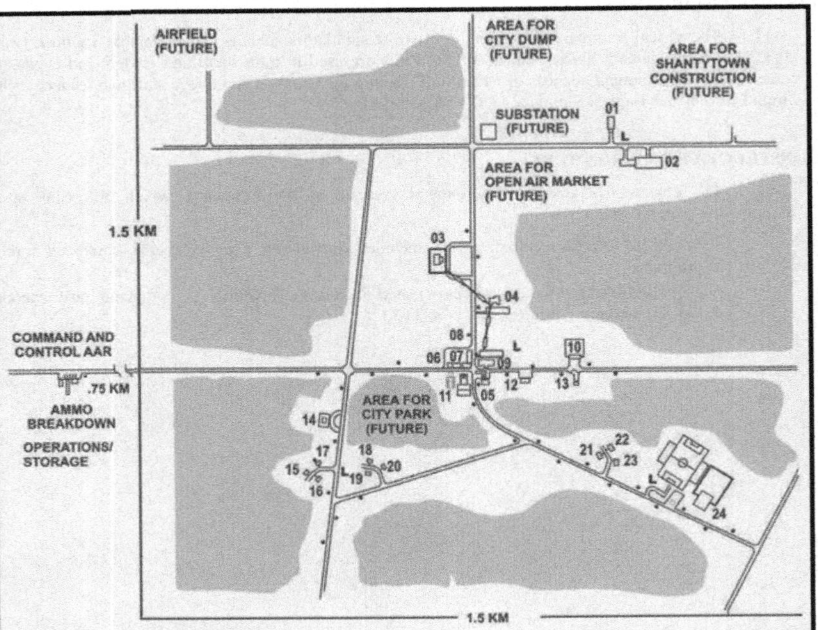

BLDG #	SIZE AND TYPE
01	1 STORY WAREHOUSE
02	1 STORY WAREHOUSE
03	3 STORY OFFICE (EMBASSY)
04	2 STORY OFFICE
05	1 STORY SERVICE STATION
06	1 STORY BUSINESS
07	1 STORY BUSINESS
08	1 STORY BUSINESS
09	3 STORY HOTEL
10	1 STORY POLICE STATION/JAIL
11	CHURCH AND CEMETARY
12	1 STORY BUSINESS
13	1 STORY BANK
14	2 STORY TOWNHOUSE
15	1 STORY RESIDENT
16	1 STORY RESIDENT
17	1 STORY RESIDENT
18	1 STORY RESIDENT
19	1 STORY RESIDENT
20	1 STORY RESIDENT
21	1 STORY RESIDENT
22	1 STORY RESIDENT
23	1 STORY RESIDENT
24	2 STORY SCHOOL

LEGEND

- FENCE WALL
- WALL
- TUNNEL SYSTEM
- TREE LINE
- UTILITY POLE
- LATRINE

SCALE: 1" = 500'

Figure 1-2. Combined Arms Collective Training Facility.

MISSION-ESSENTIAL TASK LIST

1-13. FMs 7-0 and 7-1 explain how to determine a unit's mission-essential task list (METL) and battle tasks, and how to plan, resource, and execute training events. Because training facilities are limited, leaders must use all available time and resources for training and maintaining combat readiness.

- System-specific FMs and TCs teach doctrine, conduct of operations, and employment of weapon systems.
- Tactics, techniques, and procedures (TTP) are lessons learned from training and combat.

1-14. The standards for evaluating training are accessible at the Digital Training Management System (DTMS), https://dtms.army.mil/DTMS.

BATTALION INTEGRATED TRAINING PLAN

1-15. The commander uses the battalion integrated training plan to set goals for training density. He focuses on combined arms operations, emphasizing dismounted Infantry maneuver. Tanks, BFVs, Strykers, attack aviation, and artillery normally provide supporting fires for the dismounted elements. UO training in the CACTF, Shoot House, and Breach Facility should be conducted semiannually and quarterly for the UAC. Army training calendars are divided into two cycles of training: *mission* and *support*. Figure 1-3 and Figure 1-4 show how the urban training strategy and facility usage might be programmed into unit cycles.

RISK ASSESSMENT

1-16. Risk assessment means making operations safe without compromising the level of realistic training (FM 7-1). Commanders continuously assess the risk of training conditions to prevent unnecessary loss of Soldiers and equipment. The degree of risk depends on conditions. Just as this principle applies to other training and warfighting, it also applies to UO. In a training situation, for example, commanders determine whether Soldiers and leaders have conducted this type of training before, whether conducting initial training at night or during limited visibility, and whether Soldiers are tired or exhausted. Risk assessment ensures UO training is not only tough and realistic but also safe. Training enhancers, such as live fire, blank ammunition, simulators, fires, smoke, and other pyrotechnics, must be fully integrated into the training to achieve the training objectives. Soldiers are required to wear earplugs, protective vests, and protective eyewear in order to protect themselves from noise and debris.

1-17. The commander considers risk and safety in urban operations training situations such as--.

- Soldiers engaging in close combat urban operations need to be safety conscious when using blanks, booby traps, and practice grenades. They may be unaware of the exact location of the OPFOR and may be susceptible to OPFOR flash suppressors, booby traps, and practice hand grenades.
- Soldiers should use ladders to enter buildings at upper stories of buildings to prevent unnecessary risks.

1-18. The commander is the safety officer and is responsible for ensuring that his unit, both leaders and Soldiers, comply with safety regulations and the units tactical SOP. All leaders must--.

- Use mission, enemy, terrain, troops, time available, and civilians (METT-TC) factors to identify risks.
- Assess possible losses and their costs.
- Select and develop risk-reduction measures.
- Implement controls by integrating them into plans and orders, SOPs, training performance standards, and rehearsals.
- Supervise and enforce risk reduction measures and safety standards at all times.

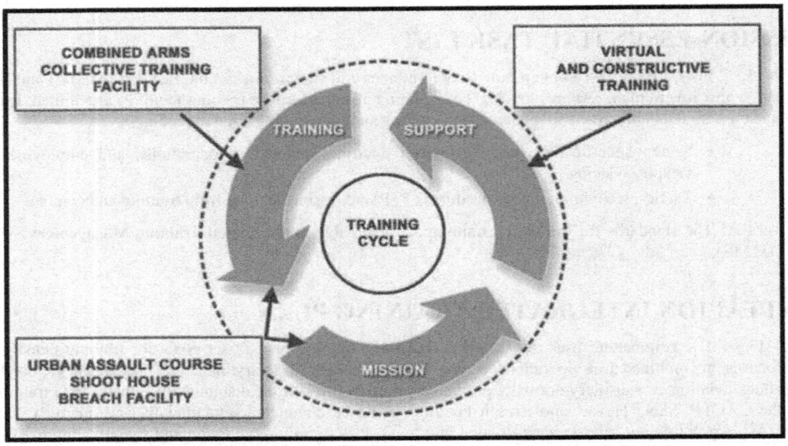

Figure 1-3. Active Army, Home Station, Urban Operations Training Strategy.

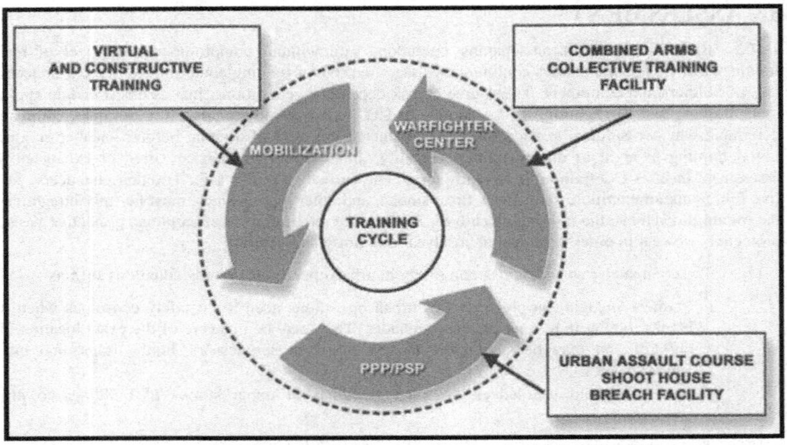

Figure 1-4. Army National Guard Urban Operations Training Strategy.

Section III. TRAINING FACILITIES

Training facilities include three that are discussed in this section: the Urban Assault Course (UAC, Chapter 2), Shoot House (SH, Chapter 3), Breach Facility (BF, Chapter 4). They also include the Combined Arms Collective Training Facility (CACTF, discussed in Chapter 5), Combat Training Centers (CTC), and any other live training that supports the UO training.

URBAN ASSAULT COURSE

1-19. The purpose of the UAC facility is to train a squad-sized force to *Enter a Building and Clear a Room*, improve grenadier gunnery, and validate subterranean TTP. However, the unit's training requirements determine the size of the unit as well as the tasks to be trained. The unit develops its organizational structure and steps for conducting the exercise based on its METL and on their training needs (Figure 1-5). (See Chapter 2 for a more detailed discussion.)

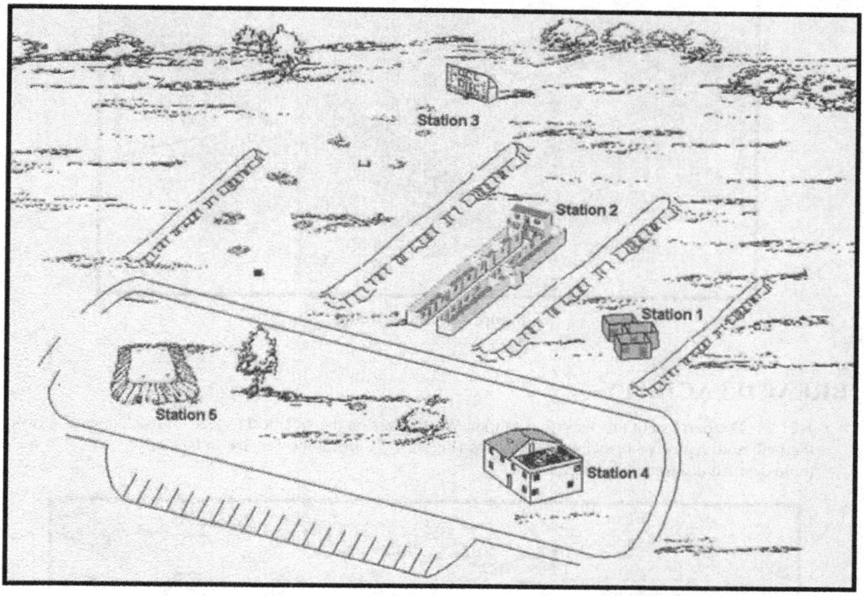

Figure 1-5. Urban Assault Course.

SHOOT HOUSE

1-20. The purpose of this facility is to provide the leader with a facility to train and evaluate fire teams, squads, and platoons during a live-fire exercise (Figure 1-6). Units are trained and evaluated on their ability to conduct tactical movement and engage targets, while practicing target discrimination in an urban environment. (See Chapter 3 for a more detailed discussion.)

Note: The specific designs of the Shoot House and the other facilities may differ from one installation to another, due to variations in training requirements.

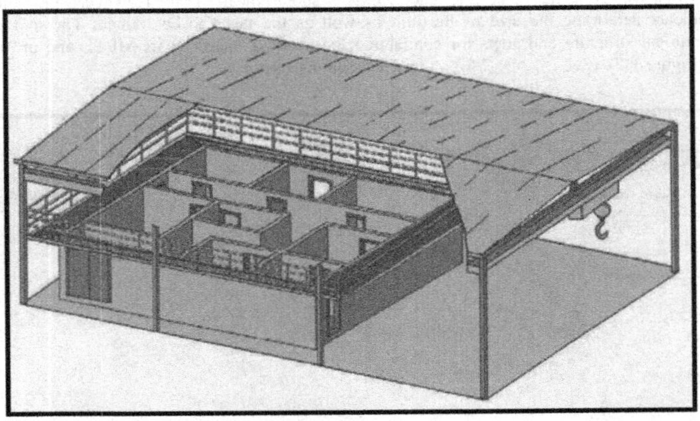

Figure 1-6. Shoot House.

BREACH FACILITY

1-21. The purpose of this facility is to train Warfighters on the technical aspects of mechanical, ballistic, thermal, and explosive breaching techniques (Figure 1-7), including for use in UO. (See Chapter 4 for a more detailed discussion.)

Figure 1-7. Breach Facility.

Section IV. CONDUCT OF TRAINING

Depending on unit proficiency, every training event should follow a *crawl-walk-run* sequence, based on clearly defined task, conditions, and standards. After each training event, leaders conduct an after-action review (AAR).

CRAWL

1-22. The trainer explains each training objective and performance standards, and then talks the Soldiers through the exercise, step by step, describing what each individual or unit must do. This training often includes a demonstration.

WALK

1-23. The Soldiers slowly practice each task to standard. Trainers coach Soldiers at this stage stopping as often as needed to correct mistakes and provide feedback. Soldiers practice each task repeatedly until they can perform it to standard.

RUN

1-24. Soldiers perform each task at full speed as if they are in combat. The Soldiers need feedback during this stage as well. The following can be used to enhance realism:

- Blank ammunition.
- Special Effects Small-Arms Marking System (SESAMS).
- Targetry.
- Pyrotechnics.
- Tactical engagement systems (MILES).
- Opposing forces (force on force).
- Nuclear, biological, and chemical simulation/weapons of mass destruction.
- Training in limited visibility conditions.

AFTER-ACTION REVIEW

1-25. Leaders conduct an after-action review (AAR) after each training event to help Soldiers and units improve their skills by providing immediate feedback. An AAR increases the benefits gained from each training exercise by allowing the leaders and the Soldiers to work together to analyze the performance of each task. Being involved in this analysis enhances the leaders and the Soldiers learning and retention ability. The AAR improves the performance of the leaders and Soldiers in later iterations by allowing interaction and feedback from these individuals during the debrief.

Section V. PREREQUISITE TRAINING

Prerequisite training teaches individuals, crews, units, leaders, and command and staff what they need to know before they are required to perform a task. Prerequisite training builds teamwork and individual confidence at all levels.

COMMAND AND STAFF TRAINING

1-26. The battalion battle staff and leadership must train before the battalion is required to conduct collective training. The battalion command and staff culminates their prerequisite training with a tactical exercise without troops (TEWTs) and command post exercises (CPXs). This sequence of training assists the commander and his staff develop better UO TTP and unit SOP.

LEADER TRAINING

1-27. A comprehensive leader's training program is the key to successful UO training (Appendix D). Time spent teaching and training the battalion leaders (trainers) during the planning and preparation phase will ensure success during the intermediate and culminating phases of training. Realistic training at the unit level will improve the leadership training officers and NCOs receive at the Basic Officer Leadership Course (BOLC), Captains Career Officer Course (CCOC), Basic Noncommissioned Officer Course [BNCOC], and Advanced Noncommissioned Officer Course [ANCOC]. Recommended training should include remedial training on those tasks that the unit had difficulty in during the previous training and evaluation.

INDIVIDUAL TRAINING

1-28. Many individual specific tasks and some collective tasks can be taught and practiced in garrison while most collective tasks are best conducted at the installation UO facilities and weapons ranges. The unit's preliminary individual training tasks should focus on three general areas: urban combat skills, urban marksmanship skills, and physical fitness. The following are example preliminary training tasks or TTP:

- Individual movement techniques.
- Battle drills.
- Weapon positioning in an urban area.
- Roadblock and vehicle search procedures.
- Dismounted and mounted urban navigation.
- Urban scanning, quick-fire, and assault fire techniques.
- Hasty urban firing positions.
- Preparation of urban firing positions.

1-29. Prior to urban collective training cycle, trainers should develop urban-specific firing conditions in conjunction with weapons firing ranges. Build facades on existing ranges, or use the unit's local UAC (if available). Urban combat skills include--

- Firing positions behind rubble, around corners, from windows or rooftops.
- Urban targets behind rubble, in windows and doorways.
- Fire commands.
- Lifting and shifting fires from lower to upper story windows.
- Fire control and fire discipline.
- Rapid firing engagements.
- ROE firing scenarios.

1-30. Urban physical fitness skills include--

- Endurance runs and upper body conditioning.
- Vaulting exercises.

- Urban-specific obstacle course negotiation.
- Leadership Reaction Course negotiation.

INDIVIDUAL TASK TRAINING

1-31. Urban operations require the Soldier to be proficient in those individual tasks that are unique to urban combat. Units should incorporate into their training regime those urban fighting TTP that have proven successful based on lessons learned from combat operations.

TASK FORCE SOLDIER--WARRIOR COMMON TASKS, SKILL LEVELS 1-3:

- 071-326-0541, *Perform Movement Techniques during an Urban Operation.*
- 071-326-0542, *Enter a Building during an Urban Operation.*
- 071-326-0550, *Prepare Positions for Individual and Crew-Served Weapons during an Urban Operation.*
- 071-326-0557, *Select Hasty Firing Positions during an Urban Operation.*
- 071-440-0027, *Execute Mechanical Breach in an Urban Operation.*
- 071-440-0028, *Engage Targets during an Urban Operation.*
- 071-440-0029, *Conduct a Breach during an Urban Operation.*
- 071-440-0003, *Conduct an Attack on a Building by a Squad during an Urban Operation.*
- 071-440-0006, *Conduct a Defense by Squad during an Urban Operation.*

FM 3-06.11, *COMBINED ARMS OPERATIONS IN URBAN TERRAIN*

Movement Techniques

- Crossing of a wall.
- Movement around corners.
- Movement past windows.
- Use of doorways.
- Movement parallel to buildings.
- Crossing open areas.
- Fire team employment.
- Movement between positions.
- Movement inside a building.

Entry Techniques

- Upper building levels.
- Use of ladders.
- Scaling of walls.
- Entry at lower levels.
- Hand grenades.

Firing Positions

- Hasty firing position.
- Prepared firing position.
- Target acquisition.
- Employment of snipers.

COLLECTIVE TASK TRAINING

1-32. Specific FMs teach doctrine, how to conduct operations, and how to employ weapon systems. TTP are lessons learned from the training and combat. The standards for conducting and evaluating collective training are accessible at the Digital Training Management System (DTMS), https://dtms.army.mil/DTMS. Those tasks that do not specifically address UO must be modified for use in an urban environment.

RESOURCES

1-33. Field manuals for the squad and platoon are--:

- FM 3-21.8, *Infantry Rifle Platoon and Squad.*
- FM 3-21.71, *Mechanized Infantry Platoon and Squad (BFV).*
- FM 3-21.9, *SBCT Infantry Rifle Platoon and Squad.*

1-34. Field manuals for the company are--.

- FM 3-21.10, *IBCT Rifle Company.*
- FM 3-21.11, *SBCT Rifle Company.*

1-35. Field manuals and mission training plans for the battalion are--.

- FM 3-21.20, *IBCT Infantry Battalion.*
- FM 3-21.21, *SBCT Infantry Battalion.*
- FM 3-06.11, *Combined Arms Operations in Urban Terrain.*
- TC 7-98-1, *Brigade and Battalion Stability and Support Operations.*

Chapter 2

Urban Assault Course

This chapter describes the Urban Assault Course (UAC), which incorporates doctrine and tactics from FM 3-06.11. The UAC contains five stations:

- Individual and Team Task/Technique Trainer Station 1
- Squad and Platoon Task/Technique Trainer Station 2
- Grenadier Gunnery Trainer Station 3
- Urban Offense/Defense Building Station 4
- Underground Trainer Station 5

These stations are designed for individual tasks and small-unit collective task training. Various types of organizations with a variety of missions can train at the UAC. Tactical operations in an urban environment may include combat to stability and support tasks. (Appendix C provides a SASO unit training matrix). The UAC is a dismounted training facility.

Note: The station discussions are arranged so that they may be removed from the book for ease of use.

PURPOSE

2-1. The purpose is to provide squad and platoon size units with a facility to train and evaluate urban operations tasks (Figure 2-1). Only Station 3, Grenadier Gunnery, is intended for live-fire training. It supports use of 40-mm TP and 5.56-mm service ammunition.

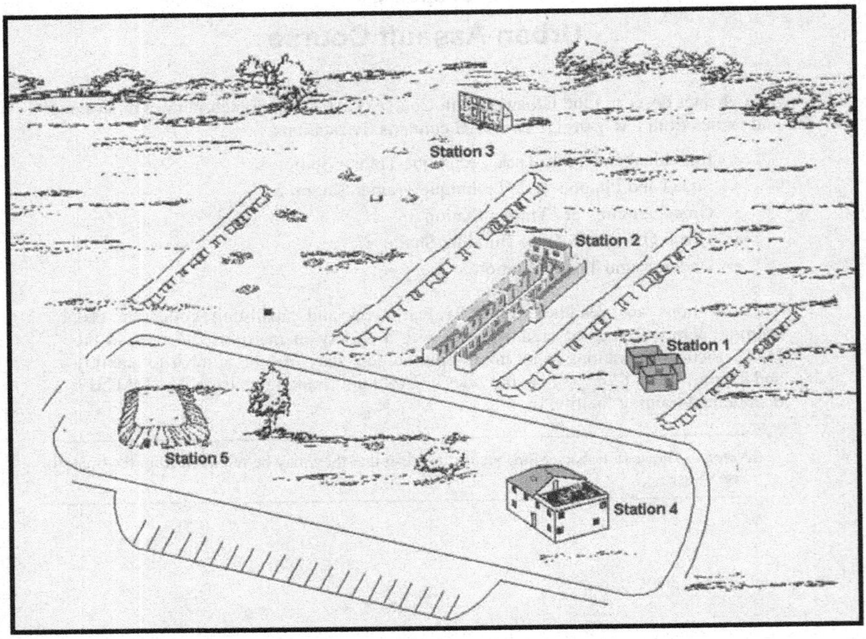

Figure 2-1. Urban Assault Course.

DESCRIPTION

2-2. The UAC has five training stations that facilitate the ***crawl-walk-run*** training concept:

Station 1, Individual and Team Trainer

2-3. This station has three rooms where team leaders and squad leaders train the basics of building and room clearing.

Station 2, Squad and Platoon Trainer

2-4. This station has four structures and multiple rooms. Squads build on tasks learned at Station 1, and begin to learn the concepts of clearing multiple buildings. This station can be used as individual buildings with a narrow street, or as rooms inside a building with a long, connecting hallway.

Station 3, Grenadier Gunnery Trainer

2-5. This is a live-fire station where M203 gunners master target engagements in an urban area, conduct tactical movement, and respond to the fire commands.

Station 4, Offense/Defense House

2-6. This is where a platoon trains to attack or defend a building. This station can also be divided into smaller training stations to reinforce training. It is also used to train tasks not yet trained at the other stations such as upper-level entry techniques.

Station 5, Underground Trainer

2-7. This station provides training for subterranean operations.

TARGETRY

2-8. Targets throughout the UAC are either 3D precision targets (for engagements less than 50 meters) or 2D nonprecision targets (for engagements greater than 50 meters).

Precision Targets

2-9. Precision targets support short distance engagement sensors. These sensors, located inside the precision targets, respond only to lethal shot placements.

- Torso lethal zone .... 18" wide x 8" high
- Neck-head lethal zone .... 4" wide x 8" high

Nonprecision Targets

2-10. Nonprecision targets support engagements at ranges greater than 50 meters. These targets are 2D pop-up type targets.

Training Audience

2-11. The facility gives Infantry, Engineer, Military Police, Cavalry, and other units a facility for training individual Soldiers through platoon-level UO skills.

Training Information

2-12. Combat units can use this facility as a mission rehearsal exercise site. Recommended training frequency is quarterly. For each UAC station, they should provide the purpose, a description, targetry, controller and safety requirements, individual UO tasks to be trained, training support requirements, and conduct of training information, if applicable:

Purpose

2-13. This UAC station [*state purpose of station*]. [*Units may use the station in the manner originally intended, or they may modify it to meet their unique training needs.*]

Description

2-14. This station is [*describe station use and design*].

Targetry

2-15. Recommended targetry for training at this station includes [*describe*]. Targets may be moved, modified, or added, depending on available target systems or training requirements.

Controller

2-16. Controller requirements for this station include [*describe*].

Safety

2-17. Safety requirements and considerations for this station include [*describe*]. [*The unit or installation range SOP may add more stringent safety requirements.*]

Individual Tasks or Skills Trained

2-18. These may be found in the Digital Training Management System located online at https://dtms.army.mil/DTMS.

Training Support

2-19. Minimum materiel requirements to support training at this station include [describe].

Conduct of Training

2-20. Example training scenarios and sequence of events for this station include [describe]. [The UAC can accommodate an assortment of training ammunition and pyrotechnics.]

STATION 1: INDIVIDUAL AND TEAM TASK/TECHNIQUE TRAINER

2-21. This station is used to train individual and collective tasks, tactics, techniques, and procedures, specifically *Enter a Building and Clear a Room* and *Engage Targets* (Figure 2-2). It has three rooms where team leaders and squad leaders train the basics of building and room clearing.

DESCRIPTION

2-22. Station 1 has a single wood structure. Its three rooms have one or more open doorways and windows, and a man-sized hole.

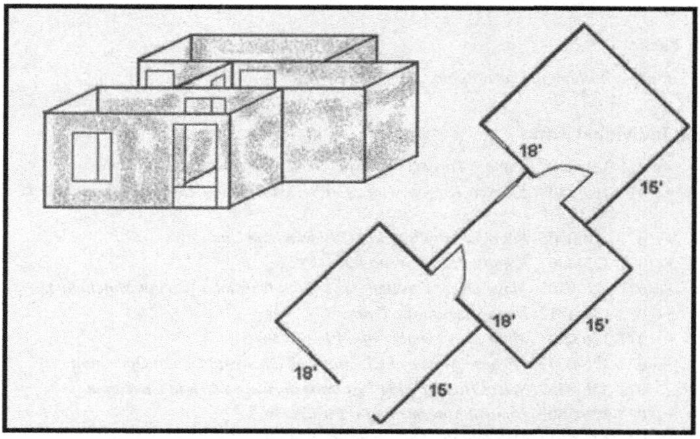

Figure 2-2. Individual and team task/technique trainer, Station 1.

TARGETRY

2-23. Targets present realistic close quarters combat situations (0 to 15 meters), which consist of a possible mixture of combatant and noncombatant targets. Target discrimination is trained by presenting up to six precision targets. Targetry sensors respond only to lethal shot placement. They also aid in training during limited visibility.

CONTROLLER REQUIREMENTS

2-24. These are based on unit requirements.

SAFETY REQUIREMENTS

2-25. Only smoke and practice grenades, or M84 stun grenades, may be used at this station.

WARNING

To avoid injury, place the weapon on SAFE when climbing through obstacles such as windows, fences, and walls during training.

INDIVIDUAL/COLLECTIVE TASKS TRAINED

2-26. This is noninclusive.

Collective Task

2-27. *Enter a Building/Clear a Room.*

Supporting Individual Tasks

- 071-311-2007, *Engage Targets with M16-Series Rifle/M4 Carbine Rifle.*
- 071-315-2308, *Engage Targets with M16-Series Rifle/M4 Carbine Using a Night Vision Sight AN/PVS-4.*
- 071-010-0006, *Engage Targets with M249 Machine Gun.*
- 071-325-4407, *Employ Hand Grenades.*
- 071-326-0503, *Move Over, Through, or Around Obstacles (except Minefields).*
- 052-193-1013, *Neutralize Booby Traps.*
- 071-326-0501, *Move as a Member of a Fire Team.*
- 071-326-0541, *Perform Movement Techniques during an Urban Operation.*
- 071-326-0557, *Select Hasty Firing Positions during an Urban Operation.*
- 071-326-5605, *Control Movement of a Fire Team.*
- 071-326-5611, *Conduct the Maneuver of a Squad.*
- 071-710-0008, *Operate Night Vision Goggles AN/PVS-7B.*

Related Tactics, Techniques, and Procedures

- FM 3-06.11, *Combined Arms Operations in Urban Terrain.*
- FM 3-22.9, *Rifle Marksmanship.*

TRAINING SUPPORT REQUIREMENTS

2-28. Use Table 2-1 for the minimum ammunition required to train at this station.

Table 2-1. Support requirements for Station 1.

Ammunition	*Rounds per Weapon*
5.56-mm Blank	10 rounds
5.56 mm Blank (Linked)	15 rounds
Practice Grenade Bodies	2 per team
Practice Grenade Fuses	2 per team

CONDUCT OF TRAINING

2-29. Table 2-2 lists the training sequence and times for Station 1.

Table 2-2. Sequence of events and times for Station 1.

Event	*Action*	*Estimated Time*
1	*Employ Hand Grenades*	30 minutes crawl, walk, run
	Perform Movement Techniques • *Engage Target with M16* • *Target Discrimination*	20 minutes crawl 20 minutes walk 10 minutes run
2	*Enter a Building*	20 minutes crawl 20 minutes walk 10 minutes run
3	*Enter and Clear a Room*	20 minutes crawl 20 minutes walk 10 minutes run
4	*Conduct an AAR*	40 minutes

Example Scenarios, Assault Team, Varied Entry Positions

2-30. Figures 2-3 and 2-4 show example scenarios. Rather than using the examples, units should tailor their training events based on their unit's METL and training needs.

Varied Entry Positions

2-31. The team leader first trains to execute *Varied Entry Positions* without targets, which is part of the *crawl* phase of training. The trainer has the assault teams position at various places to demonstrate how the start position can affect the final points of domination. (Figure 2-3).

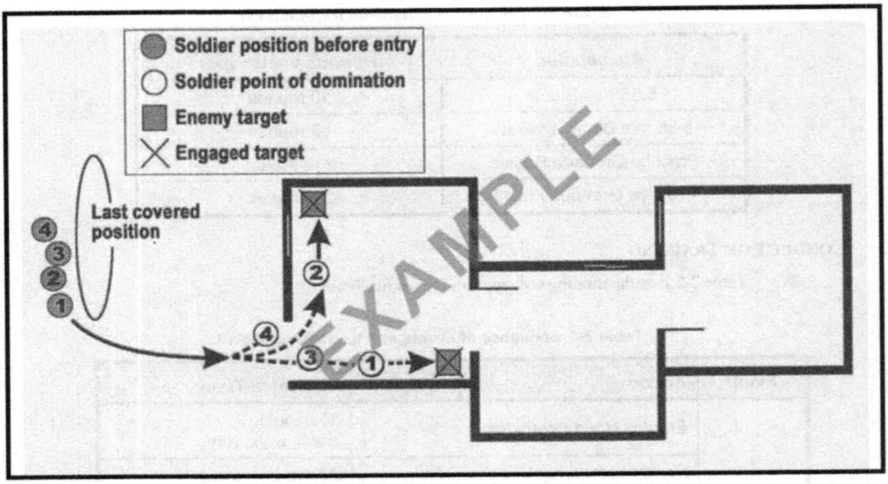

Figure 2-3. Example Station 1 scenario, *Varied Entry Positions*.

Multiple Room-Clearing Techniques

2-32. The team leader trains his Soldiers to execute *Varied Entry Positions* using targets in a multiple room scenario. This is part of the *walk* phase. The trainer has the assault teams enter the room and engage targets. He has them use using precision marksmanship techniques with the Tactical Engagement System (TES) or SESAMS (Figure 2-4). The assault team enters the first room and engages the targets. As the number two Soldier clears his area of responsibility, he sees and engages the target in the second room, but remains in place until the assault teams finishes clearing the first room. Once the room is clear, the assault team moves to the second room and continues clearing, or lets the next team move through.

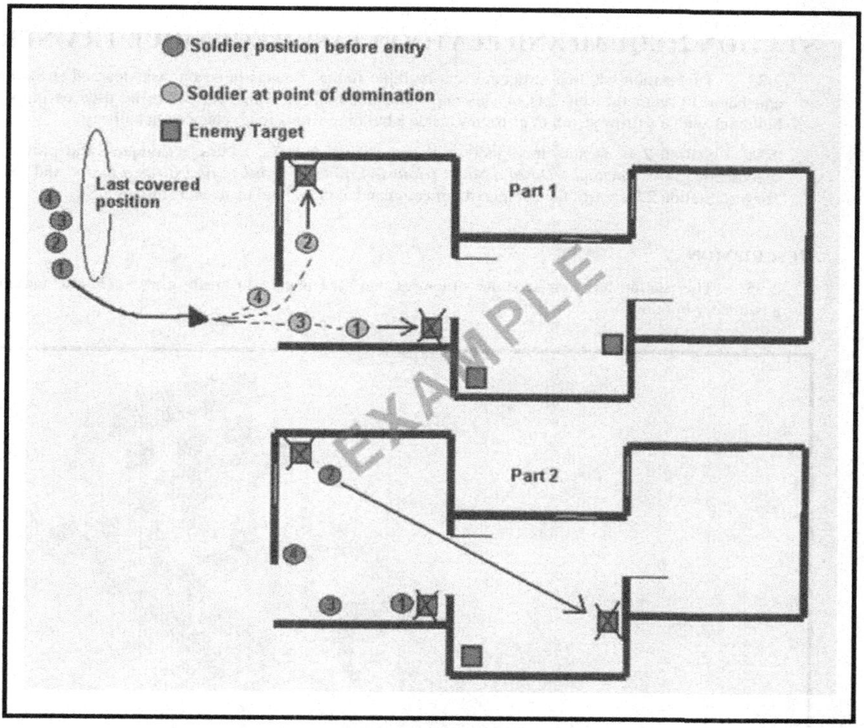

Figure 2-4. Example Station 1 scenario,
Multiple Room-Clearing Techniques.

STATION 2: SQUAD AND PLATOON TASK/TECHNIQUE TRAINER

2-33. This station has four structures and multiple rooms. Squads build on tasks learned at Station 1, and begin to learn the concepts of clearing multiple buildings. This station can be used as individual buildings with a narrow street, or as rooms inside a building with a long, connecting hallway.

2-34. Station 2 is used to train individual and collective tasks, tactics, techniques, and procedures, specifically, *Move Tactically Down a Street (Hallway)*, *Enter a Building and Clear a Room*, and *Engage Targets*. Station 2 increases the complexity of command, control, and maneuver (Figure 2-5).

DESCRIPTION

2-35. This station has four separate structures that face inward to create a street. It also introduces a two-story building.

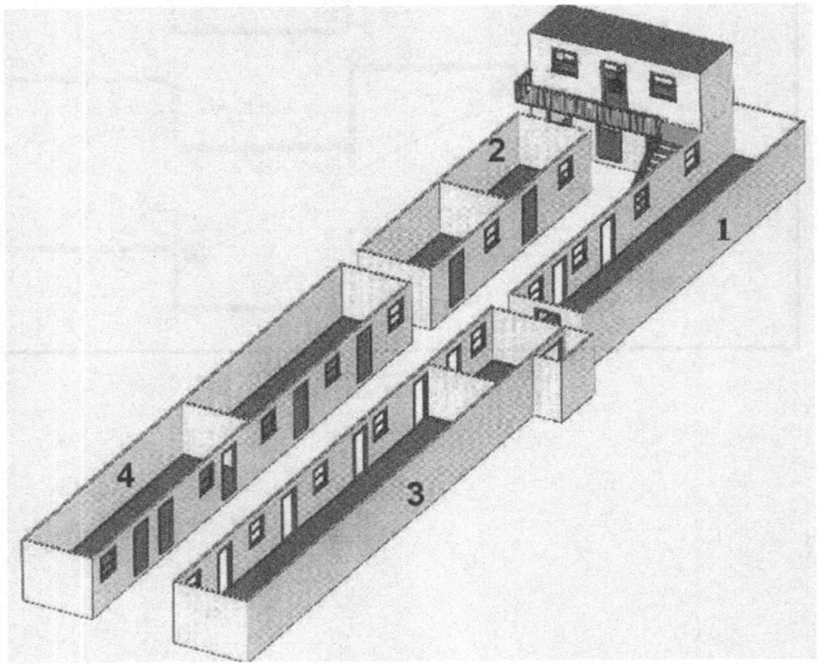

Figure 2-5. Squad and platoon task and technique trainer, Station 2.

> Building 1 is L-shaped.
>
> - The long leg of the building to the right looks like a single-story building.
> - The short leg of the building to the front has two floors. A landing along the length of the second floor has outside stairs. An inside stairway also joins the floors.
> - The two legs of the "L" are adjacent, but without interior access to each other.
>
> Building 2 has a single floor and two rooms, each with a door and window. An open area much like a street separates this structure from the first.
>
> Building 3 is a row of wooden buildings along a street. It has many windows and doorways. At the far end is an adjoining room.
>
> Building 4 is on the left. It mirrors Building 3, plus it has a wall on the side at the near end.

2-36. Units can place additional walls in any structure by using target cloth or other material. They can do this to add to the number of rooms or to increase the complexity of the training (Appendix A).

TARGETRY

2-37. Station 2 targets present realistic close quarters combat situations (0 to 15 meters), with the option to customize the mixture of combatant and noncombatant targets. Target discrimination is trained on up to ten precision targets. Targetry sensors respond only to lethal shot placement, and facilitate training during limited visibility.

CONTROLLER REQUIREMENTS

2-38. These are based on unit requirements.

SAFETY REQUIREMENTS

2-39. At this station, Soldiers may only use smoke grenades, practice grenades, and M84 stun grenades.

> **WARNING**
>
> **To avoid injury, place the weapon on SAFE when climbing through obstacles such as windows, fences, and walls during training.**

INDIVIDUAL AND COLLECTIVE TASKS TRAINED

2-40. This list is noninclusive.

Collective Task--*Enter and Clear a Building*

Related Collective Tasks

- *Conduct Consolidation and Reorganization (Company/Platoon).*
- *Treat Casualties.*
- *Evacuate Casualties.*
- *Defend MOUT Building.*

- *Conduct Tactical Movement.*
- *Enter and Clear a Room.*

Supporting Individual Tasks

- 052-193-1013, *Neutralize Booby Traps.*
- 071-010-0006, *Engage Targets with M249 Machine Gun.*
- 071-311-2007, *Engage Targets with M16-Series Rifle/M4 Carbine Rifle.*
- 071-315-2308, *Engage Targets with M16-Series Rifle/M4 Carbine Using a Night Vision Sight AN/PVS-4.*
- 071-325-4407, *Employ Hand Grenades.*
- 071-326-0501, *Move as a Member of a Fire Team.*
- 071-326-0503, *Move Over, Through, or Around Obstacles (except Minefields).*
- 071-326-0541, *Perform Movement Techniques during an Urban Operation.*
- 071-326-0557, *Select Hasty Firing Positions during an Urban Operation.*
- 071-326-0608, *Use Visual Signaling Techniques.*
- 071-326-5605, *Control Movement of a Fire Team.*
- 071-326-5611, *Conduct the Maneuver of a Squad.*
- 071-710-0008, *Operate Night Vision Goggles AN/PVS-7B.*
- 113-571-1022, *Perform Voice Communications.*
- 191-377-5250, *Handle Enemy Personnel and Equipment.*

Related Tactics, Techniques, and Procedures

- FM 3-06.11, *Combined Arms Operations in Urban Terrain.*
  -- Chapter 3, *Offensive Operation.*
  -- Chapter 5, *Fundamental Combat Skills.*
  -- Appendix M, *Field-Expedient Breaching of Common Urban Barriers.*
- FM 3-22.9, *Rifle Marksmanship.*

TRAINING SUPPORT REQUIREMENTS

2-41. Table 2-3 shows the minimum ammunition required to train at this station.

Table 2-3. Support requirements for Station 2.

Ammunition	Rounds per Weapon
5.56-mm blank	16
5.56-mm blank (linked)	24
Practice grenade fuses	8 per squad
Practice grenade bodies	8

CONDUCT OF TRAINING

2-42. Table 2-4 shows the sequence of events and times for Station 2. Figure 2-6 through Figure 2-28 show example scenarios. However, rather than use these examples, the unit should customize its training events based on its METL and training needs.

Table 2-4. Sequence of events and times for Station 2.

Event	*Action*	*Estimated Time*
1	*Conduct Tactical Movement in a Built-Up Area* *Move Across an Open Area* *Move Parallel to Buildings*	10 min Crawl 20 min Walk 10 min Run
2	*Enter and Clear a Room (Squad)* *Clear a Stairwell* *Clear a Hallway* *Conduct a Breach*	10 min Crawl 20 min Walk 10 min Run
3	*Enter and Clear a Building (Platoon)* *Clear Multiple Rooms* *Consolidate and Reorganize*	10 min Crawl 20 min Walk 10 min Run
4	*Conduct AAR (Squad)*	10 min
NOTE: Training should include techniques outlined FM 3-06.11.		

2-43. The squad is clearing Station 2 as if it were a single building, and the street a hallway. Fire Team A moves down the left side, throws the appropriate grenade IAW ROE, enters, engages the target, and clears the room on that side. The last Soldier remains outside and secures the rear Fire Team B prepares to enter the next room (Figure 2-6).

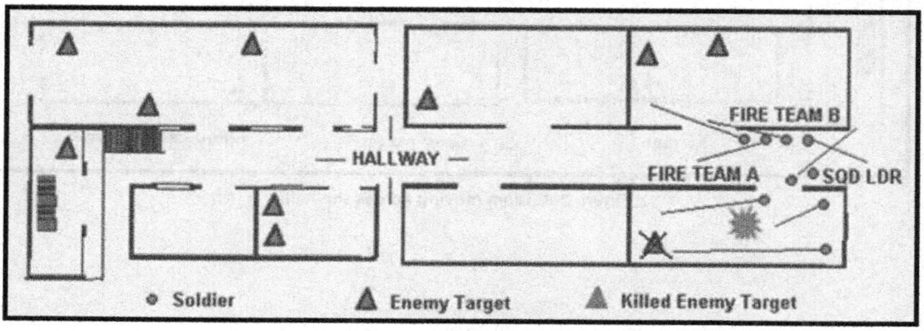

Figure 2-6. Conducting tactical movement.

2-44. Fire Team B throws the appropriate grenade, enters, engages the targets, and clears the room on the right while the last Soldier secures the rear Fire Team A prepares to continue movement to the next room (Figure 2-7).

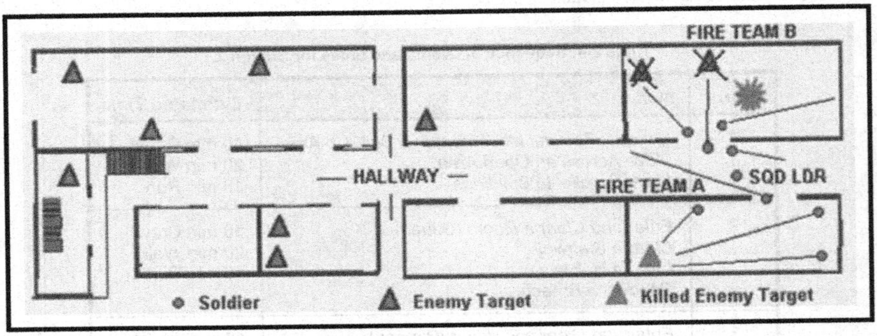

Figure 2-7. Team B--Clearing a building.

2-45. Fire Team A moves across the hall, throws the appropriate grenade, enters, engages the target, and clears the next room Three team members enter the room, the fourth remains in the doorway and secures the hallway (Figure 2-8).

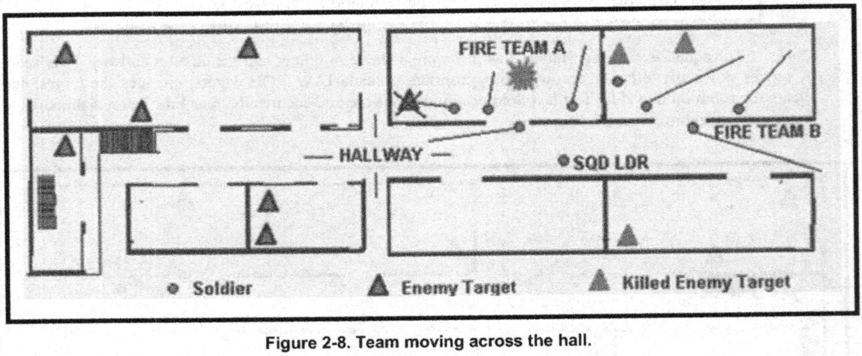

Figure 2-8. Team moving across the hall.

2-46. Fire Team B moves in a cross-cover formation down the hallway to the next room, throws the appropriate grenade, and clears the room. As Fire Team B passes the security man of Team A, he turns and becomes rear security. The last Soldier in Fire Team B remains by the door and secures the hallway (Figure 2-9).

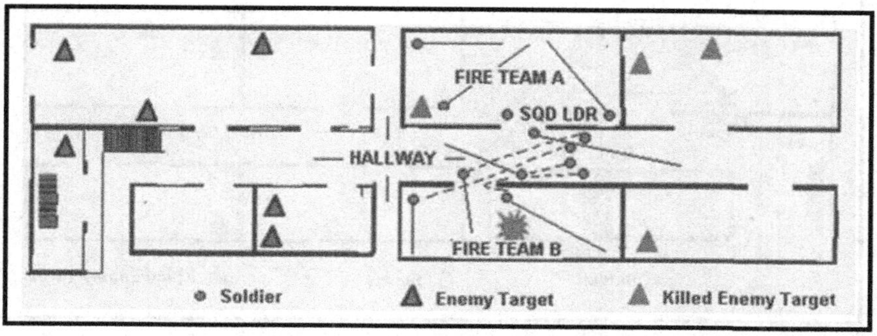

Figure 2-9. Team B--Moving in cross cover formation.

2-47. Fire Team A clears the hallway intersection, while Team B uses the Rolling-T technique to pass through Team A en route to the next room (Figure 2-10).

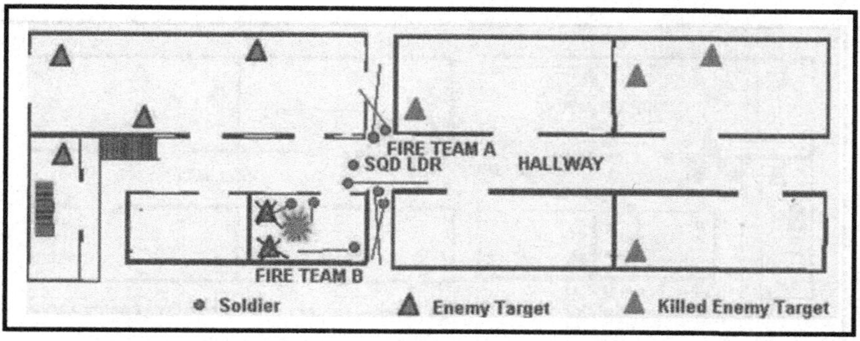

Figure 2-10. Team A--Clearing a hallway intersection.

2-48. Fire Team B throws the appropriate grenade, enters, engages the targets, and clears the room. Only three Soldiers enter, while the fourth Soldier provides rear security. Fire Team A prepares to move (Figure 2-11).

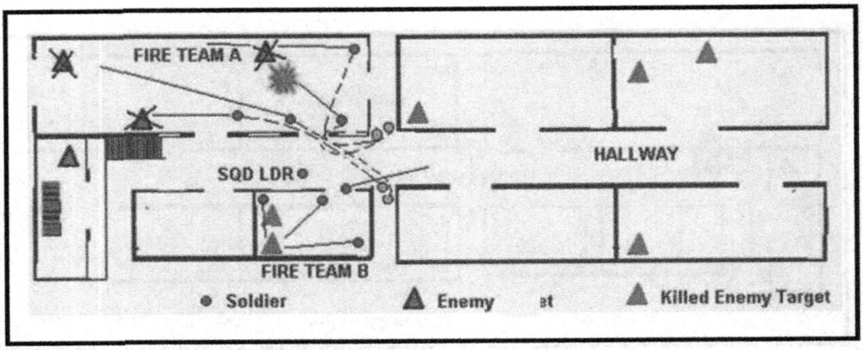

Figure 2-11. Team B--Clearing a room.

2-49. Fire Team A moves from the intersection and throws the appropriate grenade. The team enters, engages the targets, and then clears the next room. All team members enter, because of the size of the room. Fire Team B prepares to move (Figure 2-12).

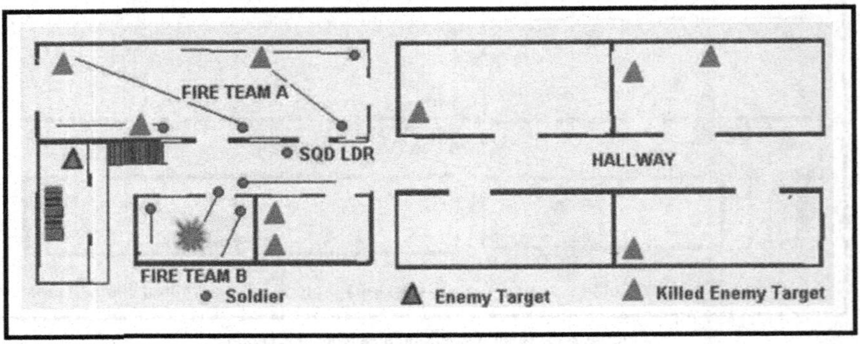

Figure 2-12. Team A--Clearing room by room.

2-50. Fire Team B employs the appropriate grenade, enters, and engages the target. Three Soldiers clear the next room. The fourth remains in the doorway and secures the rear. Fire Team A prepares to move up the stairs (Figure 2-13).

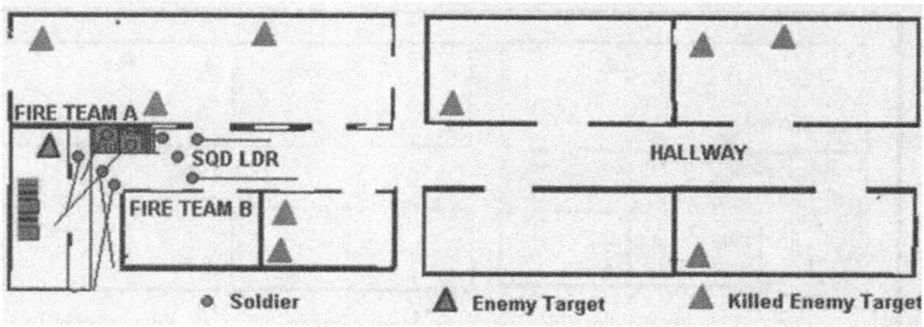

Figure 2-13. Team B--Clearing the next room.

2-51. Fire Team A moves up the stairs and onto the landing. The team prepares to enter the room off the landing. Fire Team B splits, and two members move around the corner and secure the area across from the stairs on the lower level. The other two members of Team B secure the hallway the squad came from (Figure 2-14).

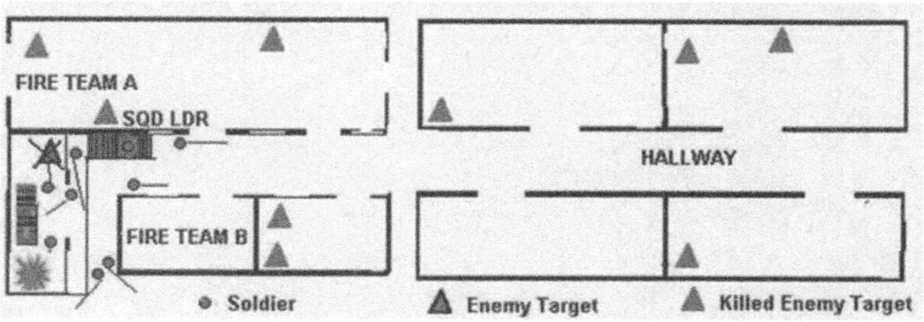

Figure 2-14. Securing the stairs and hallway.

2-52. Fire Team A members employ the appropriate grenade, enters the room, engages the target, and clears it. They find another set of stairs, and prepare to move up them. Fire Team B continues to secure the rear. The other team members move to the end of the short hallway and secure both directions (Figure 2-15).

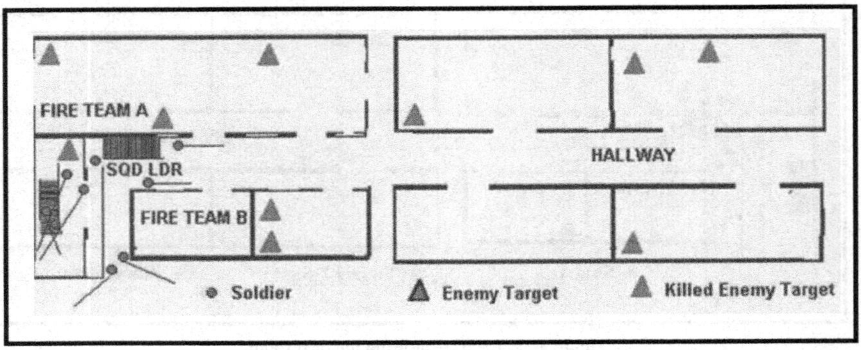

Figure 2-15. Preparing to clear more stairs.

2-53. Fire Team A clears the stairs and the landing. Fire team B continues to secure the ground level, ready to assist Team A, if necessary (Figure 2-16).

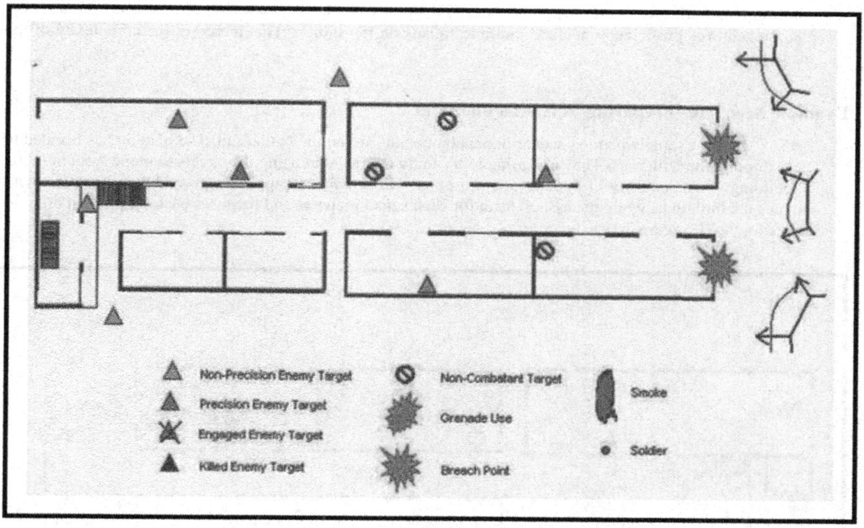

Figure 2-16. Team A--Clearing stairs and landing.

2-54. The previous example used no noncombatant targets, but used blanks/SESAMS and TES. This technique is the *crawl* phase for this station. Once the squad is ready to move on to more challenging exercises, the platoon leader/sergeant can add targets into the scenario. As proficiency increases, mixed combatant and noncombatant targets can be added. At any point, the platoon leader/sergeant can, until the platoon achieves proficiency, include multiple squads on the station. This increases the complexity of the scenarios.

Example Scenario--Preparing to Gain a Foothold

2-55. In this example platoon scenario, noncombatants are on the battlefield. The platoon has decided to clear through the buildings. They are going to try to do this without using the streets to move from building to building. Two squads will assault, and the other will remain in support/reserve. The support element isolates the buildings, while the assault team for each squad prepares and breaches each building in order to gain a foothold (Figure 2-17).

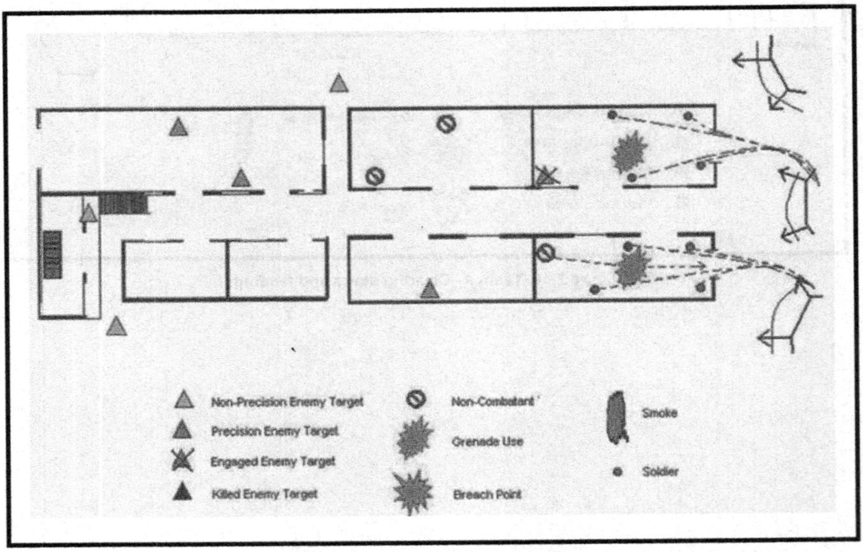

Figure 2-17. Preparing to gain a foothold.

2-56. The assault teams throw a grenade into the breach and their members follow, engaging enemy targets. They secure and move noncombatants to the collection point (CP) (Figure 2-18).

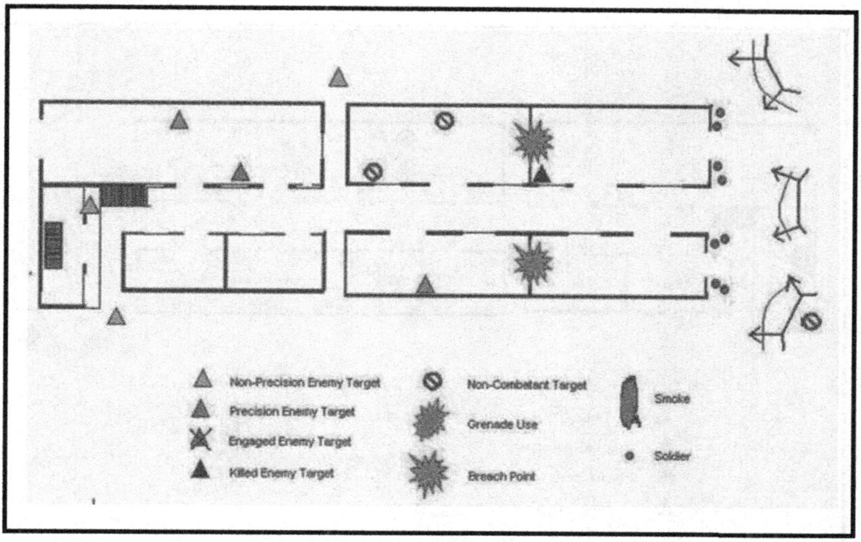

Figure 2-18. Securing and moving noncombatants to the command post.

2-57. The teams prepare and detonate wall breaches to continue movement through the building. Once the charges are ready for detonation, the assault teams move to a covered position and execute the breach (Figure 2-19).

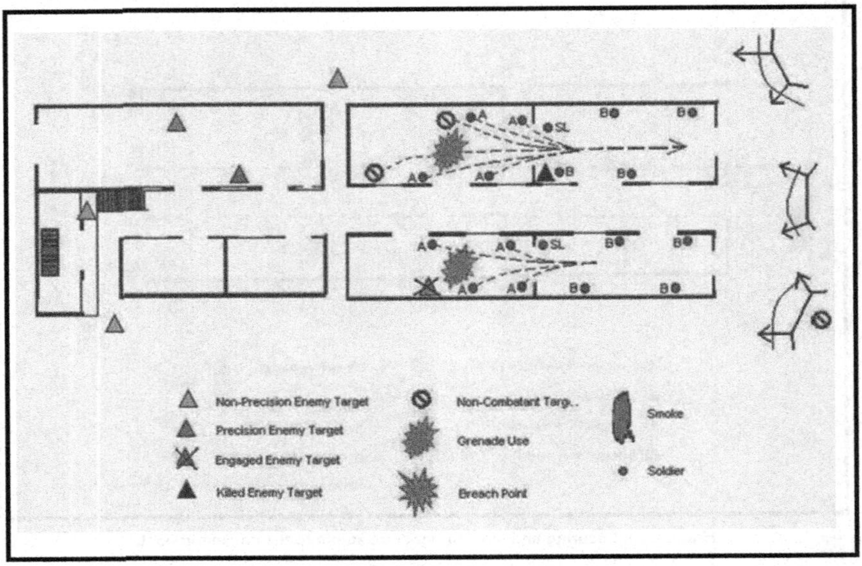

Figure 2-19. Executing breach.

2-58. The assault team employs grenades into the next room, enters the room, eliminates the enemy, and then secures and moves the noncombatants to a designated safe area. At the same time, the next assault team moves into the foothold (Figure 2-20).

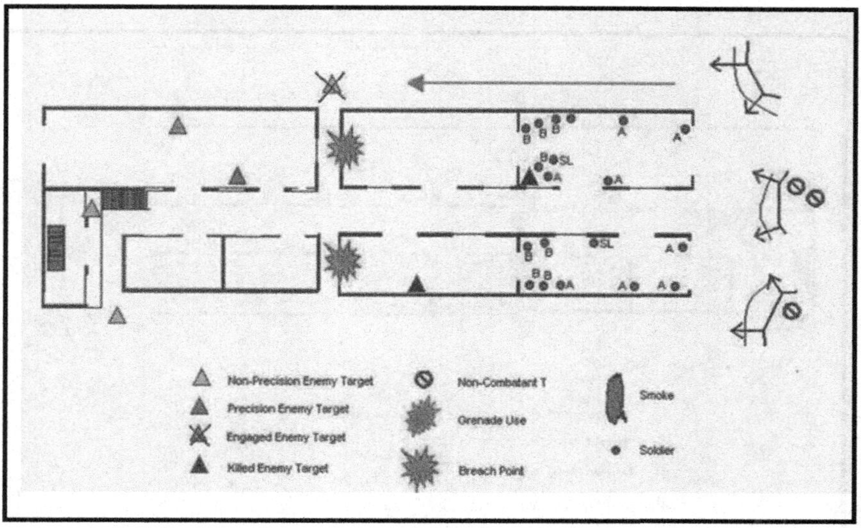

Figure 2-20. Moving to the foothold.

2-59. With the second room secured, another assault team moves forward and prepares to breach the wall Once the charges are placed, the teams move back to a covered position before detonating the charge. With the breach a success, the first nonprecision target is exposed and engaged by the support element (Figure 2-21).

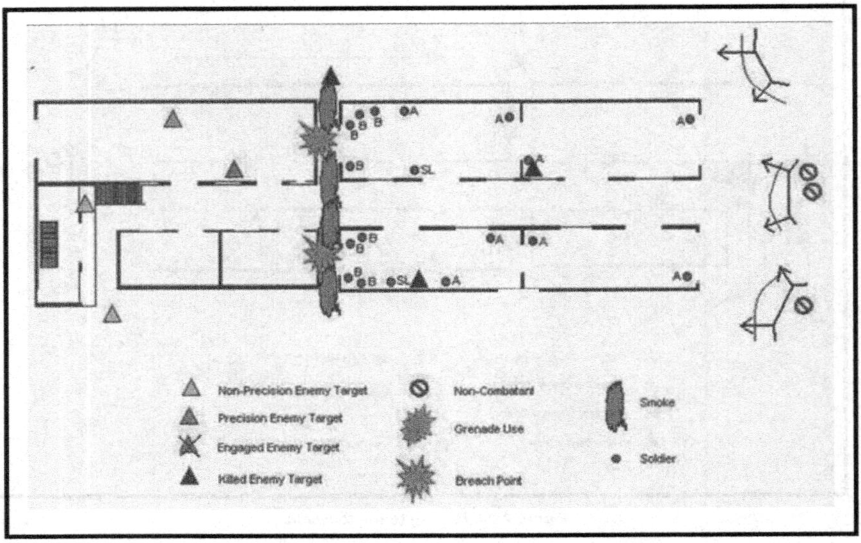

Figure 2-21. Breaching the wall.

2-60. Upon inspecting the breach, the assault teams find that they have an alley with another wall to breach. They employ smoke to obscure the teams placing the breaching charge. The first assault teams take up securing positions throughout the already cleared areas. Follow-on assault teams move to a covered position and execute the breach (Figure 2-22).

Note: Use smoke only in the alley, not inside the buildings.

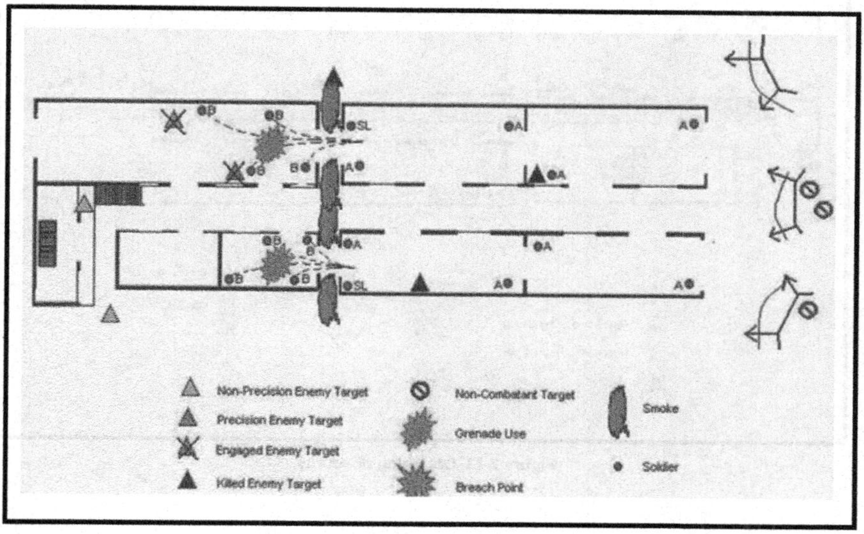

Figure 2-22. Securing a wall in an alley.

2-61. Once the assault teams execute the breach, they must determine if they need to use more smoke to cover their movement. Either way, once they have enough smoke to cover their movement, they throw in a grenade, enter, engage enemy targets, and clear the rooms (Figure 2-23).

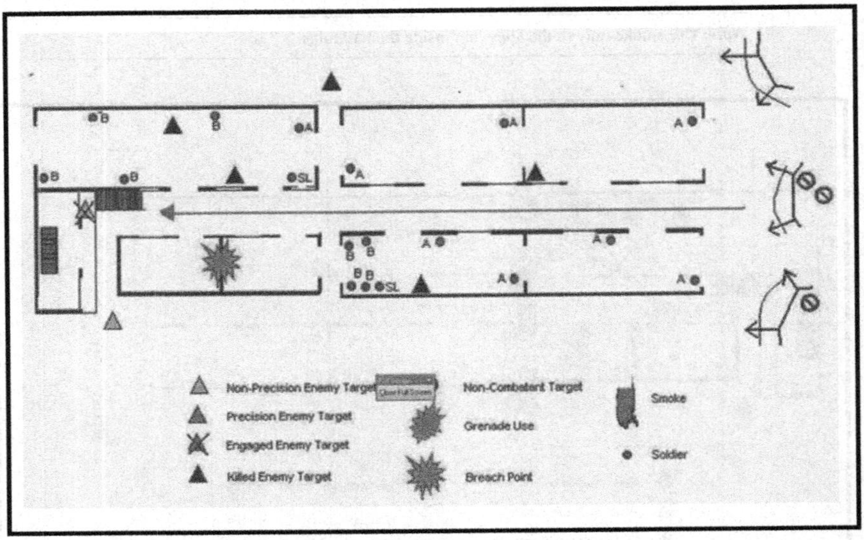

Figure 2-23. Obtaining obscurity.

2-62. While one team finds itself at the end of the building in their sector, the other team is faced with another wall and more clearing. One team prepares to breach the wall, and the other team takes up a hasty position and prepares for a counterattack. When the teams execute the breach, the next nonprecision target presents, and the support element engages it (Figure 2-24).

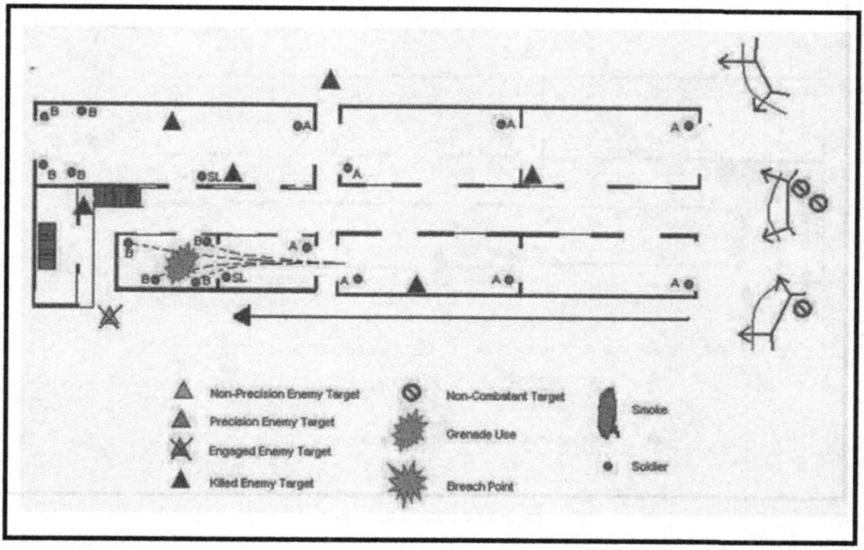

Figure 2-24. Clearing the wall.

2-63. Someone throws a grenade through the breach. The assault team enters and secures the room. When the grenade explodes inside the room, the final nonprecision target presents, and the support element engages it (Figure 2-25).

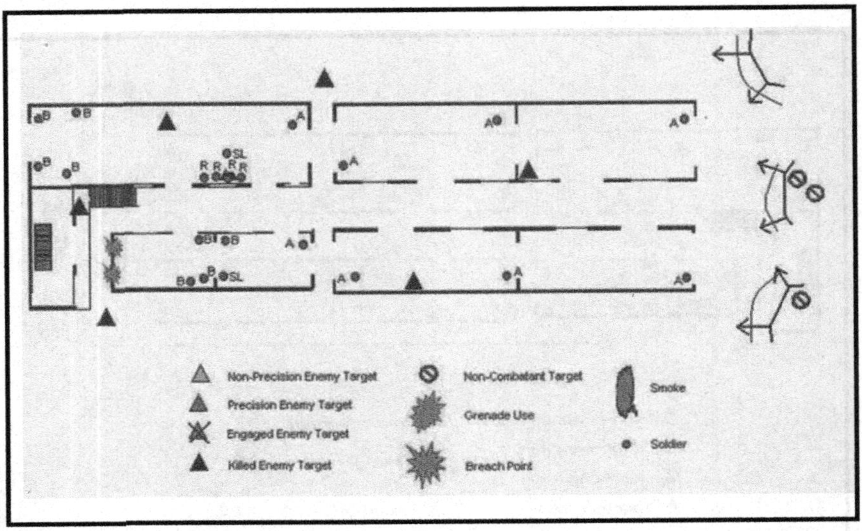

Figure 2-25. Securing a room.

2-64. With the first squad reaching its limit of advance, and with the second squad reaching the end of the building, the platoon leader calls another assault team forward from the reserve. The reserve assault team moves forward through the buildings, and positions itself to clear up the outside stairs of the two-story structure. The second squad places and detonates loophole charges on the far wall of the building. This supports the clearing of the last building (Figure 2-26).

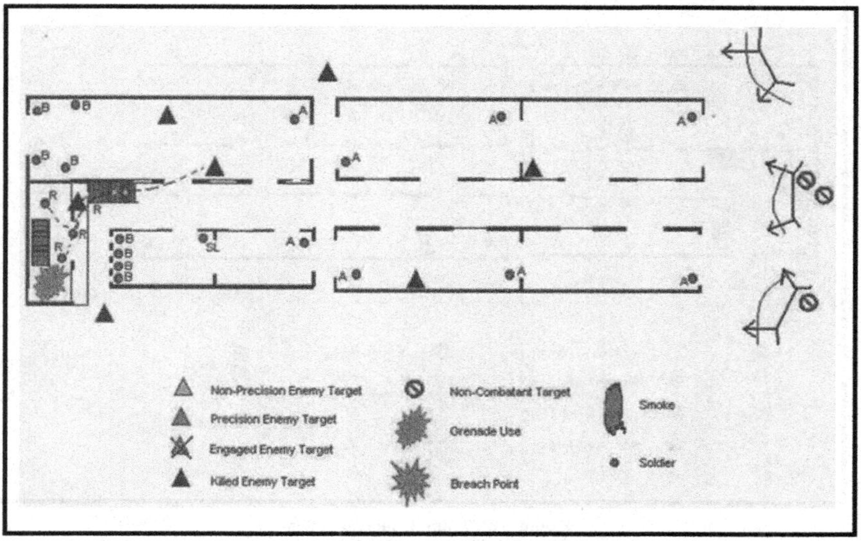

Figure 2-26. Reserve assault team moving forward.

2-65. Team B of the second squad moves into supporting fire positions. The new assault teams move up the outside stairs to the second floor of the final structure. They clear the balcony and the first room on the second floor. The support-by-fire position shift fires to the lower floor of the same structure (Figure 2-27).

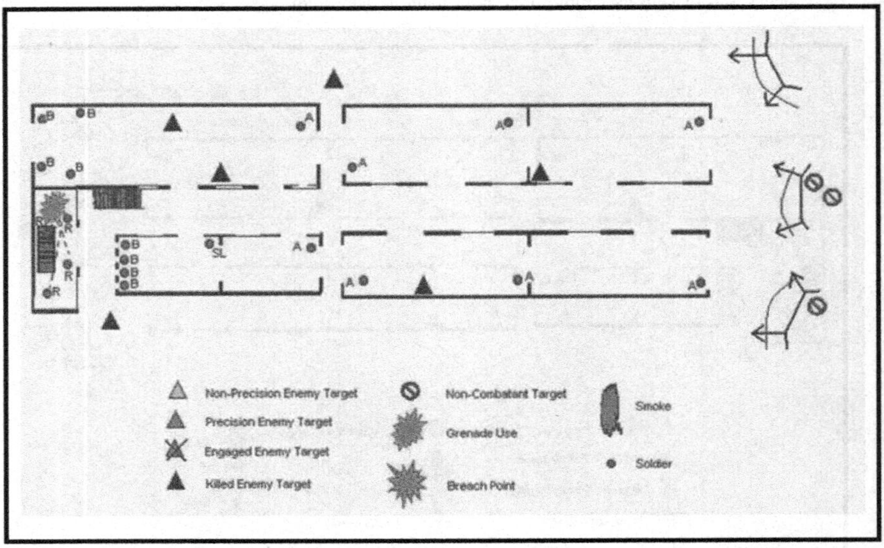

Figure 2-27. Team B, second squad--moving into the supporting fire position.

2-66. Once the floor is secure, the assault team moves down the stairs and clears the remaining floor (Figure 2-28). Meanwhile, the platoon repositions, and then consolidates and reorganizes.

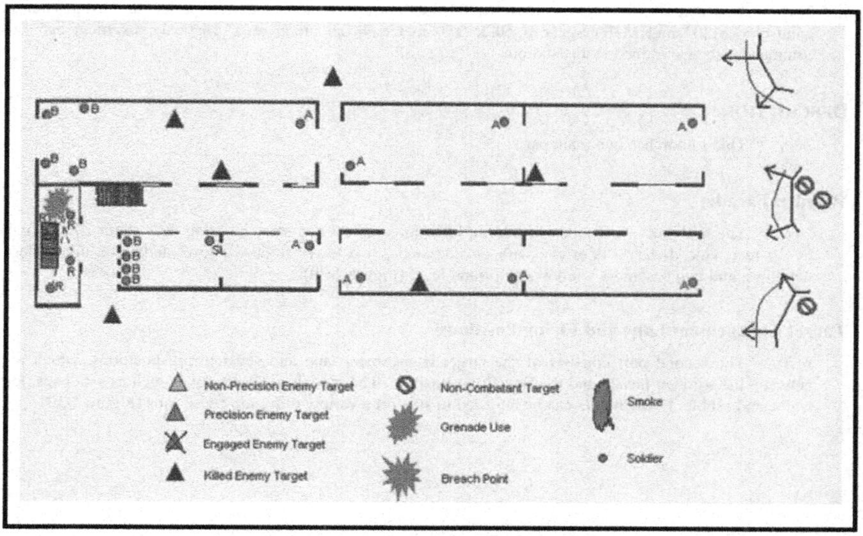

Figure 2-28. Consolidation and reorganization.

STATION 3: GRENADIER GUNNERY TRAINER

2-67. This is a live-fire station where M203 dual-purpose weapon gunners master the engagement of point-type M203 and M16 targets at 50 to 150 meters in an urban area. Tactical movement and fire commands are also trained at this station.

DESCRIPTION

2-68. This station has two main parts:

Wooden Facade

2-69. The first part is a wooden facade wall that looks like a two-story building. It is about 20 feet high by 30 feet wide. It has a doorway with outside steps, two lower floor windows and three upper floor windows, and two basement windows at ground level (Figure 2-29).

Target Engagement Lane and Firing Positions

2-70. The second part consists of the target engagement lane and seven firing positions, which are between the wooden façade and the first firing position. These positions have cover such as sandbags, log walls, and rubble. Firing points can be adjusted to support a variety of training scenarios (Figure 2-30).

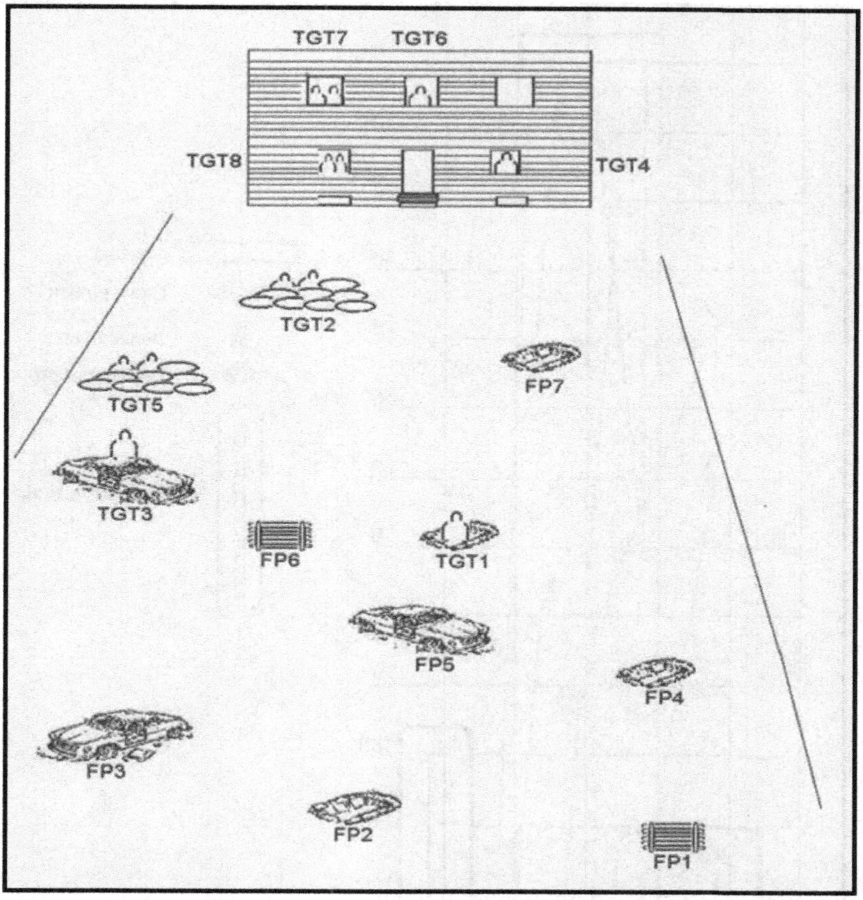

Figure 2-29. Part 1, Grenadier Gunnery Trainer, Station 3, wooden facade.

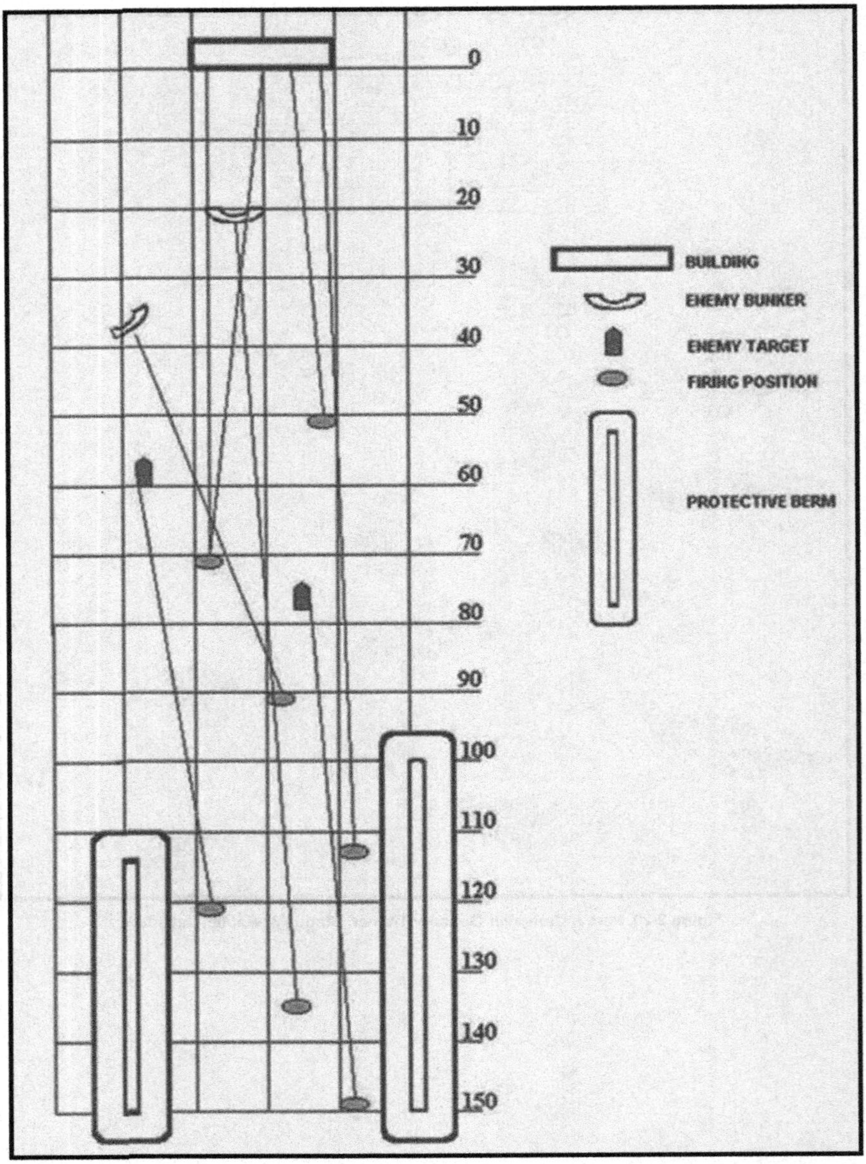

Figure 2-30. Part 2, Grenadier Gunnery Trainer, Station 3, target engagement lane and firing positions.

WARNING

Never engage targets with the 40-mm at ranges less than 31 meters, which is the minimum safe range. Trainers ensure the station accommodates this requirement as well as the 14-meter arming range.

TARGETRY

2-71. This station has twelve target mechanisms. These mechanisms are arranged so as to influence the gunner's decision as to which munition to use.

CONTROLLER REQUIREMENTS

2-72. As required by unit.

SAFETY REQUIREMENTS

2-73. Check the unit SOP and local safety regulations regarding live-fire ranges. Ensure all weapon systems are on SAFE before moving to the next firing position. Ensure that no 40-mm target is engaged at less than minimum arming range.

INDIVIDUAL AND COLLECTIVE TASKS TRAINED

2-74. These lists are noninclusive:

Supporting Individual Tasks

- 071-311-2007, *Engage Targets with M16-Series Rifle/M4 Carbine Rifle.*
- 071-311-2130, *Engage Targets with M203 Grenade Launcher.*
- 071-326-0503, *Move Over, Through, or Around Obstacles (except Minefields).*
- 071-326-0541, *Perform Movement Techniques during an Urban Operation.*
- 071-326-0557, *Select Hasty Firing Positions during an Urban Operation.*
- 071-326-0608, *Use Visual Signaling Techniques.*

Related Tactics, Techniques and Procedures

- FM 3-06.11, *Combined Arms Operations in Urban Terrain.*
  -- Chapter 3, *Offensive Operations.*
  -- Chapter 5, Fundamental Combat Skills.
- Appendix M, *Field Expedient Breaching of Common Urban Barriers.*

Training Support Requirements

2-75. Table 2-5 shows the minimum materials required to train at this station.

Table 2-5. Support requirements for Station 3.

Ammunition	*Rounds for each Weapon*	*Targets Required*
5.56-mm Ball	8	Four nonprecision human target mechanisms.
40-mm TP	8	Eight nonprecision human target mechanisms.

Conduct of Training

2-76. Whereas Table 2-5 showed the minimum amount of ammunition required for this station, Table 2-6 shows an example sequence of events and times, and Table 2-7 shows an example target scenario.

Table 2-6. Sequence of events and times for Station 3.

Event	*Action*	*Estimated Time*
1	*Perform Movement Techniques* *Select Hasty Firing Position* *Engage Targets With 5.56-mm (dry run)* *Engage targets with 40mm*	20 min
2	*Perform Movement Techniques* *Select Hasty Firing Position* *Engage Targets with 5.56mm (live-fire)* *Engage targets with 40mm*	10 min
NOTE: Training should include techniques outlined in FM 3-06.11		

Table 2-7. Example target scenario (presentation order) for Station 3.

Firing Position	*Weapon*	*Presentation Time*	*Range to Target*	*Target Presented*
1	5.56mm	6 sec	67 meters	Single Human Target (TGT 1)
2	40mm	2 min	105 meters	Multiple Human Targets (TGT 2)
3	5.56mm	6 sec	63 meters	Single Human Target (TGT 3)
4	5.56mm	6 sec	105 meters	Single Human Target (TGT 4)
5	40mm	2 min	57 meters	Multiple Human Targets (TGT 5)
6	40mm	2 min	70 meters	Multiple Human Targets (TGT 6)
6	5.56mm	6 sec	70 meters	Single Human Target (TGT 8)
7	40mm	2 min	50 meters	Multiple Human Targets (TGT 7)

STATION 4: URBAN OFFENSE/DEFENSE BUILDING, TASK/TECHNIQUE TRAINER

2-77. This station trains to attack or defend a building. Collective and individual TTP are trained. The platoon practices the task steps and performance measures for offensive and defensive tasks. This station increases the complexity of command, control, and maneuver (Figure 2-31).

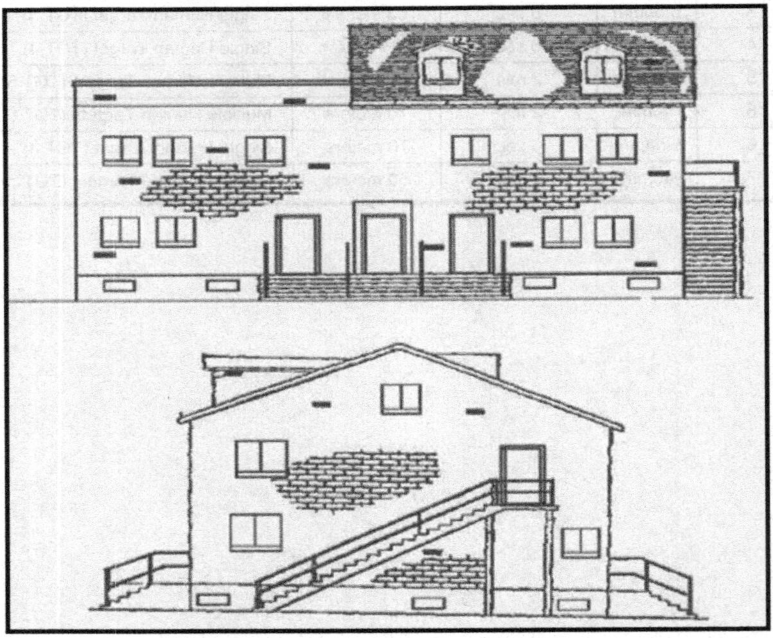

Figure 2-31. Urban offense/defense building, Station 4.

DESCRIPTION

2-78. This station is a two-story building with a basement. Each floor has several rooms, windows, doorways, loopholes, and mouseholes. The gabled roof has windows and a hatch to the attic below and half of the rooftop is flat. The building has exterior and interior stairways.

TARGETRY

2-79. This station has ten target mechanisms.

CONTROLLER REQUIREMENTS

2-80. These are based on unit requirements.

SAFETY REQUIREMENTS

2-81. Only smoke, practice, and M84 stun grenades may be used at this station. It is not a live-fire station. Training is conducted using TES with blank 5.56-mm, or TES or SESAMS. Units should consult with their supporting combat engineers for the proper demolition effects simulators (DES) to use against the blow panels throughout the station.

DANGER

TO AVOID ASPHYXIATION, NEVER DETONATE SMOKE GRENADES INSIDE BUILDINGS. THEY ARE *ALL* DANGEROUS IN AN ENCLOSED SPACE.

WARNINGS

1. **To avoid injury, place your weapon on SAFE when climbing through obstacles such as windows, fences, and walls.**
2. **To avoid inflicting carbon burns or more serious injury, never fire blank ammunition at other Soldiers within 1 meter.**

DEMOLITION EFFECTS SIMULATORS

2-82. Demolition effects simulators (DES) replicate breaching to gain entry to buildings and rooms. The walls of the UAC, Shoot House, and CACTF facilities may be breached using wall breach charge DES. Generally, the interior and exterior walls in these facilities are made with blow panels, plasterboard, or similar material, and measure about 4 feet square.

2-83. The DES charges for use at the UAC, shoot house, and the CACTF are constructed to breach a predetermined entry point without damaging the facility.

Field-Expedient Charges

- Flex-linear shaped charge.
- Doorknob charge.
- E-silhouette charge.

Operational Charges

2-84. Unit trainers must consult with installation safety and combat engineers to determine the appropriate reduced charge. See also FM 3-06.11.

INDIVIDUAL OR COLLECTIVE TASKS TRAINED

Collective Task

- FM 3-21.8 (7-8).
  -- 07-3-9018, *Clear a Building (Platoon/Squad).*
  -- 07-6-1216, *Defend an Urban Area (Battalion/BCT).*

Related Collective Tasks

- 07-2-5027, *Conduct Consolidation and Reorganization (Company/Platoon).*
- 08-2-0003. *Treat Casualties.*
- 08-2-0004. *Evacuate Casualties.*
- 19-3-3100, *Process Detainees at a Collecting Point (CP).*
- 07-2-1342, *Conduct Tactical Movement (Company/Platoon).*
- 07-2-1189, *Conduct a Dismounted Tactical Road March (Company/Platoon).*
- 07-2-1198, *Conduct a Mounted Tactical Road March (Company/Platoon).*
- Warrior Battle Drill 12, *Enter and Clear a Room.*

Supporting Individual Tasks

- 052-193-1013, *Neutralize Booby Traps.*
- 071-010-0006, *Engage Targets with M249 Machine Gun.*
- 071-311-2007, *Engage Targets with M16-Series Rifle/M4 Carbine Rifle.*
- 071-315-2308, *Engage Targets with M16-Series Rifle/M4 Carbine Using a Night Vision Sight AN/PVS-4.*
- 071-325-4407, *Employ Hand Grenades.*
- 071-326-0501, *Move as a Member of a Fire Team.*
- 071-326-0503, *Move Over, Through, or Around Obstacles (except Minefields).*
- 071-326-0541, *Perform Movement Techniques during an Urban Operation.*
- 071-326-0557, *Select Hasty Firing Positions during an Urban Operation.*
- 071-326-0608, *Use Visual Signaling Techniques.*
- 071-326-5605, *Control Movement of a Fire Team.*
- 071-326-5611, *Conduct the Maneuver of a Squad.*
- 071-710-0008, *Operate Night Vision Goggles AN/PVS-7B.*
- 113-571-1022, *Perform Voice Communications.*
- 191-377-5250, *Handle Enemy Personnel and Equipment.*

Related Tactics, Techniques, and Procedures

- FM 3-06.11, *Combined Arms Operations in Urban Terrain.*
  -- Chapter 3, *Offensive Operations.*
  -- Chapter 5, *Fundamental Combat Skills.*
  -- Appendix M, *Field Expedient Breaching of Common Urban Barriers*
- FM 3-22.9, *Rifle Marksmanship.*

Training Support Requirements

2-85. Table 2-8 shows the minimum ammunition required to train at this station.

Table 2-8. Ammunition requirements for Station 4.

Ammunition	Rounds for each Weapon
5.56-mm Blank	16
5.55-mm Blank (Linked)	24
7.62-mm Blank	100
Practice Grenade Bodies	5
Practice Grenade Fuzes	10 per platoon
M84 Stun Grenade Bodies	5
M84 Stun Grenade Fuzes	10 per platoon
Smoke Grenades	4 per platoon

Conduct of Training

2-86. Table 2-9 shows the sequence of training for Station 4. Figures 2-32 and 2-33 show example scenarios of training events that can be conducted at this station. These are only examples. The unit should develop training events based on its METL and training needs.

Table 2-9. Sequence of events and times for Station 4.

Event	Action	Estimated Time
1	*Clear a Building*	2 Hours Crawl
		1 Hour Walk
		30 Minutes Run
2	*Defend Built-up Area/Building*	1 Hours Crawl
		1 Hour Walk
		2. Hour Run
3	*Conduct AAR*	1 Hour

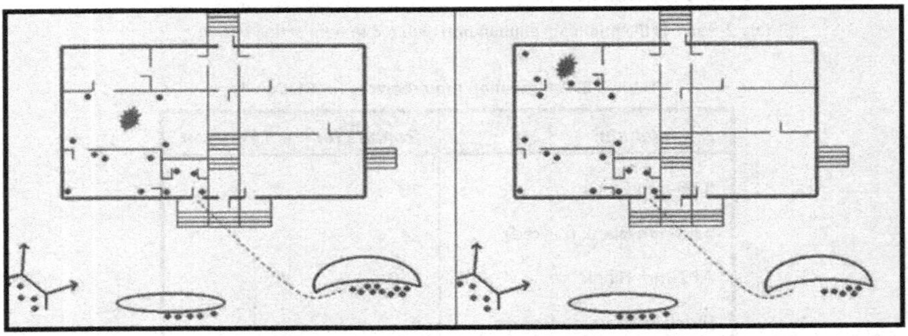

Figure 2-32. Event 1, *Clear a Building*.

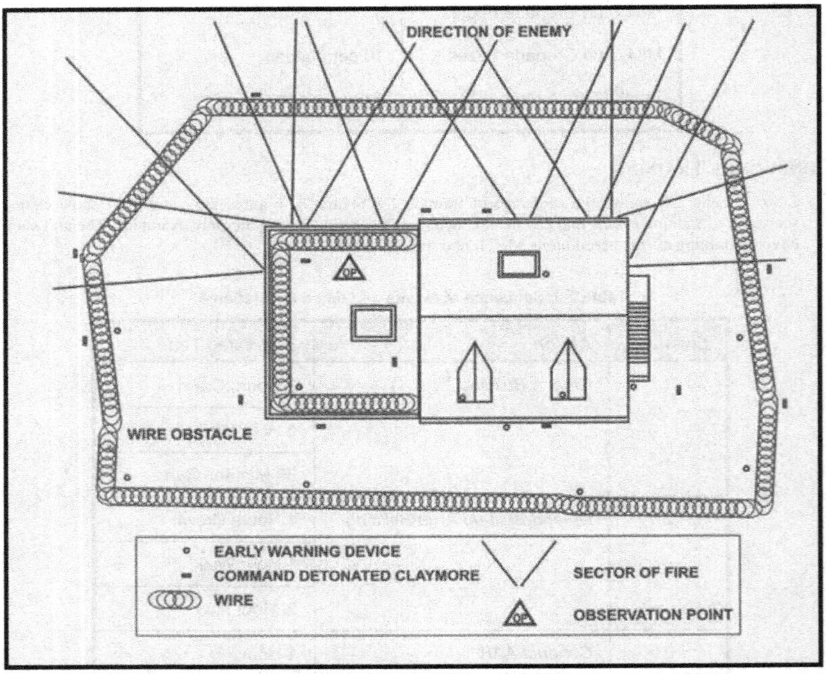

Figure 2-33. Event 2, *Defend a Building*.

STATION 5: UNDERGROUND TRAINER

2-87. This station provides training up to squad level for subterranean operations, specifically, the task *Clear and Move Tactically in a Subterranean Environment.*

DESCRIPTION

2-88. This station is a sewer system built in an irregular pattern, with four manhole covers. Two access tunnels lead from the outside tunnel to the underground loop, and the station is completely covered by dirt, only the manholes and ground level entrances are exposed (Figure 2-34).

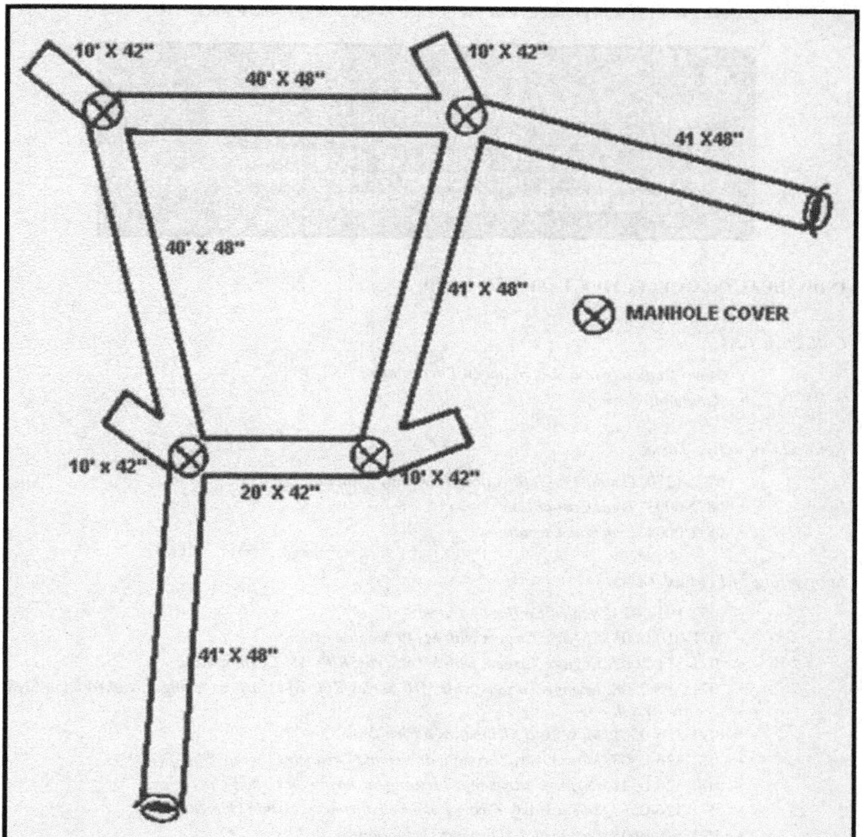

Figure 2-34. Underground Trainer, Station 5.

TARGETRY

2-89. Station 5 is a force-on-force training station and has no targetry.

CONTROLLER REQUIREMENTS

2-90. As required by unit.

SAFETY REQUIREMENTS

2-91. Soldiers should wear their helmets at all times when in the sewer system to avoid head injuries. Smoke grenades or M84 stun grenades *will not* be used in the underground trainer.

DANGER

TO AVOID ASPHYXIATION WHEN IN THE TRAINER, NEVER DETONATE SMOKE GRENADES IN THE SEWER. USING ANY TYPE OF EXPLOSIVE SIMULATOR IS DANGEROUS IN AN ENCLOSED SPACE. THE PROTECTIVE MASK IS NOT EFFECTIVE IN OXYGEN-DEFICIENT ATMOSPHERES.

INDIVIDUAL OR COLLECTIVE TASKS TRAINED

Collective Task

- *Move Tactically in a Subterranean Environment.*
- *Reconnoiter Area.*

Related Collective Tasks

- 07-2-1270, *Conduct an Infiltration (Company/Platoon).*
- 08-2-0003. *Treat Casualties.*
- 08-2-0004. *Evacuate Casualties.*

Supporting Individual Tasks

- 052-193-1013, *Neutralize Booby Traps.*
- 071-010-0006, *Engage Targets with M249 Machine Gun.*
- 071-311-2007, *Engage Targets with M16-Series Rifle/M4 Carbine Rifle.*
- 071-315-2308, *Engage Targets with M16-Series Rifle/M4 Carbine Using a Night Vision Sight AN/PVS-4.*
- 071-326-0501, *Move as a Member of a Fire Team.*
- 071-326-0503, *Move Over, Through, or Around Obstacles (except Minefields).*
- 071-326-0541, *Perform Movement Techniques during an Urban Operation.*
- 071-326-0557, *Select Hasty Firing Positions during an Urban Operation.*
- 071-326-0608, *Use Visual Signaling Techniques.*
- 071-326-5605, *Control Movement of a Fire Team.*
- 071-326-5611, *Conduct the Maneuver of a Squad.*
- 071-710-0008, *Operate Night Vision Goggles AN/PVS-7B.*

- 113-571-1022, *Perform Voice Communications.*
- 191-377-5250, *Handle Enemy Personnel and Equipment.*

Related Tactics, Techniques and Procedure

- FM 3-06.11, *Combined Arms Operations in Urban Terrain.*
  -- Chapter 3, *Offensive Operations.*
  -- Chapter 5, *Fundamental Combat Skills.*
  -- Appendix M, Field Expedient Breaching of Common Urban Barriers.
- FM 3-22.9, *Rifle Marksmanship.*

TRAINING SUPPORT REQUIREMENTS

2-92. Use Table 2-10 for the minimum ammunition required to train at this station.

Table 2-10. Ammunition requirements for Station 5.

Ammunition	*Rounds for each Weapon*
5.56-mm Blank	20
5.55-mm Blank (Linked)	40

CONDUCT OF TRAINING

2-93. Table 2-11 shows the sequence of training for this station. These are only examples. The unit should develop training events based on its METL and training needs.

Table 2-11. Sequence of events and times for Station 5.

Event	*Action*	*Estimated Time*
1	*Reconnoiter*	30 minutes (Crawl) 30 minutes (Walk) 30 minutes (Run)
2	*Conduct an Infiltration (Co/Plt)*	30minutes (Crawl) 30 minutes (Walk) 30 minutes (Run)
3	*Conduct AAR*	30 minutes

Chapter 3
Shoot House

This chapter describes the live-fire Shoot House and provides layouts and training scenarios. Fighting in built-up areas is fragmented; gains are often small. Therefore, units depend on the initiative, skill, and discipline of small-unit leaders and individual Soldiers. Leaders must be competent and confident in their urban operations skills and must have the courage to accomplish their missions while isolated from their parent units. Soldiers and leaders require mission-type orders that allow for decentralized execution.

PURPOSE

3-1. The Shoot House provides the leader with a facility to train and evaluate the unit during a live-fire exercise. Units are trained and evaluated on their ability to conduct tactical movement, engage targets, conduct breaches, and practice target discrimination in an urban environment.

DESCRIPTION

3-2. This facility has two parts, the live-fire facility and the AAR room:

LIVE-FIRE FACILITY

3-1. The live-fire facility is a single-level structure with a barn roof, catwalk, and eight adjoining rooms. Each room has video capability for AARs. The walls are bulletproof and prevent ricochets. Each side has an entrance. Replaceable blow panels throughout the facility can withstand reduced DES blasts. These also offer alternative entry points. A crane system in the roof allows trainers to selectively replace doorways and blow panels with solid wall sections. Before training explosive and mechanical breaching operations, trainers should place mobile exterior baffles or containment walls placed by each entry to prevent rounds and demolition effects from escaping. All of these capabilities let the unit vary its training scenarios. For safety, the Shoot House is windowless (Figure 3-1).

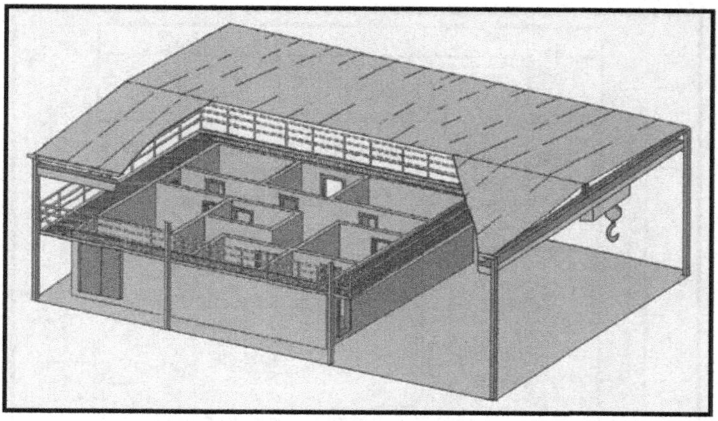

Figure 3-1. Shoot House.

AAR BUILDING

3-2. The AAR building has a conference/AAR room that can accommodate up to a platoon size element, an instrumentation and targetry control room, and an AAR edit and projection room that can handle a formal AAR (Figure 3-2).

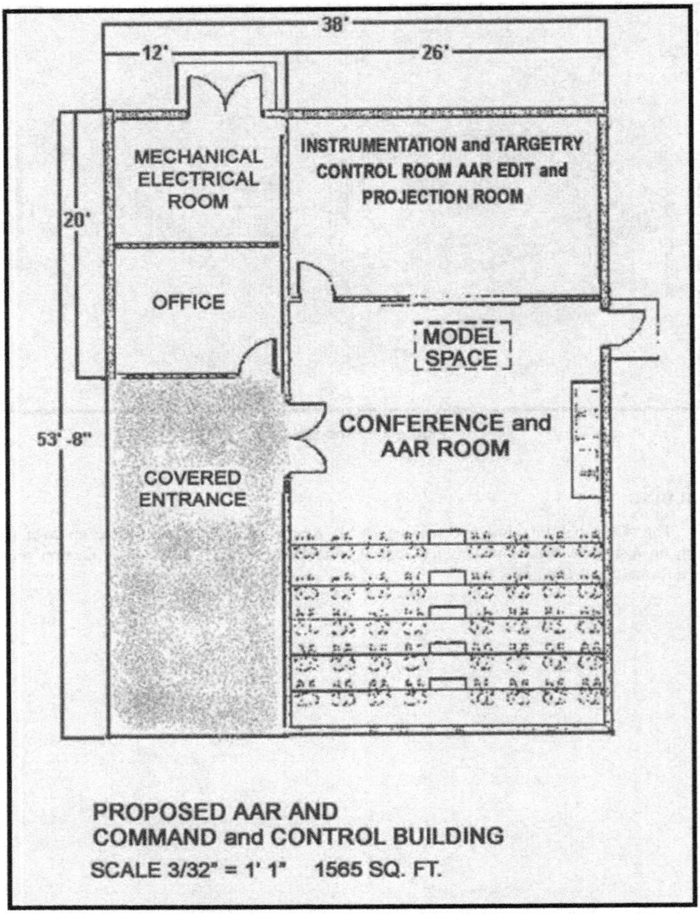

Figure 3-2. Command and control building and AAR room.

TRAINING AUDIENCE

3-3. The facility provides leaders the ability to train and evaluate the unit during a live-fire exercise. However, the unit's training requirements determine the size of the unit and the tasks to be trained at this facility. The unit develops its own organization and the steps for conducting the exercise based on its METL and training needs.

TRAINING INFORMATION

3-4. Tactical operations in an urban environment may range from combat missions to stability and support missions. Commanders should note that the Shoot House *is not* suited for the employment of live-fragmentation/concussion grenades, 40-mm HE/HEDP, antiarmor weapons, mortars, artillery, or air delivered munitions. Crew-served weapon systems, M2 and M1 tanks, and machine guns can be employed against supplemental targetry that is exterior to the Shoot House. The Shoot House has specially constructed areas that are used to conduct either mechanical or explosive breaching techniques. When conducting explosive breaching, units should consult with local Engineers for appropriate DES charges and consult Appendix A before conducting the training.

CONDUCT OF TRAINING

3-5. The unit develops realistic time frames for each operation or task. The times are based on the *crawl-walk-run* concept. The training event or STX can be arranged in any order as long as they follow a tactical scenario. *Assault a Building, Enter a Building and Clear a Room*, and *Clear a Building* are the primary collective tasks trained at this facility.

EXECUTION

3-6. Tables 3-1 through 3-5, pages 3-4 through 3-7, shows a sequence of events and times, and Figure 3-3 through Figure 3-6, pages 3-6 through 3-9, show example scenarios. These are only examples. The unit should develop training events based on its METL and training needs. However, the leader should do the following when conducting training at the live-fire Shoot House:

1. Identify the unit's critical missions (see commander's training guidance), and select those tasks that support specific tasks related to urban operations.
2. Arrange the related tasks in the order of their importance to mission accomplishment.
3. Assess the unit's current level of proficiency (strengths and weaknesses), and determine the degree of difficulty of each task. The training events and scenarios should start from the least difficult to the most difficult tasks. This gives the Soldiers confidence.
4. Identify the individual and leader tasks to be trained.
5. Conduct individual training first, because it is the first step in the building-block approach. It is also the basis for unit or collective task training. Mastery of individual skills is a precondition for effective training.
6. Set up training conditions based on unit proficiency. At first, keep the conditions basic, until Soldiers attain proficiency. Increase difficulty and realism as training progresses. Then, add limited visibility, CBRN factors, and live fires with service or short-range training ammunition (SRTA).
7. Understand that the Shoot House is a live-fire capable training facility. However, it can be used to train any level of proficiency, including blank, TES, SESAM, SRTA, and ball (no tracers). The leader assesses the unit's proficiency.

Table 3-1. Sequence of events and times for the platoon.

Element	Event action	Estimated Time
All 1st Squad 2d Squad 3rd Squad	1. Conduct precombat checks (for example; install, align, and troubleshoot MILES equipment; conduct inspection; draw equipment). 2. Support Operations. 3. Conduct Rehearsals. 4. Conduct Event 1 (Crawl).	2 Hours
3rd Squad 1st Squad 2d Squad	5. Support Operations. 6. Conduct Rehearsals. 7. Conduct Event 1 (Crawl).	2 Hours
2d Squad 3rd Squad 1st Squad	8. Support Operations. 9. Conduct Rehearsals. 10. Conduct Event 1 (Crawl).	2 Hours
1st Squad 2d Squad 3rd Squad	11. Support Operations. 12. Conduct Rehearsals. 13. Conduct Event 1 (Walk).	2 Hours
3rd Squad 1st Squad 2d Squad	14. Support Operations. 15. Conduct Rehearsals. 16. Conduct Event 1 (walk).	2 Hours
2d Squad 3rd Squad 1st Squad	17. Support Operations. 18. Conduct Rehearsals. 19. Conduct Event 1 (walk).	2 Hours
1st Squad 2d Squad 3rd Squad	20. Support Operations. 21. Conduct Rehearsals. 22. Conduct Event 1 (Run).	1 Hours
1st Squad 2d Squad 3rd Squad	23. Support Operations. 24. Conduct Rehearsals. 25. Conduct Event 1 (Run).	1 Hours
1st Squad 2d Squad	26. Support Operations. 27. Conduct Event 1 (Run).	1 Hours

Table 3-2. Event 1, Squad Sequence (*Crawl*).

Element	Event Action	Estimated Time
1 (CRAWL)	**Terminal Learning Objective** Familiarize the squad with how to enter and clear a room during the crawl phase of training **Tasks** *Enter and Clear a Room (Squad)* • *Perform Movement Technique* • *Select Hasty Firing Position* *Conduct a Breach (Mechanical)* • *Employ Hand Grenades* • *Engage Targets with M16A2/M4* • *Engage Targets with M249 Machine Gun*	2 Hours
NOTES: 1. Perform controlled pairs and target discrimination techniques (FM 3-22.9). 2. Include techniques (FM 3-06.11) and performance measures (FM 3-21.8 and DTMS).		

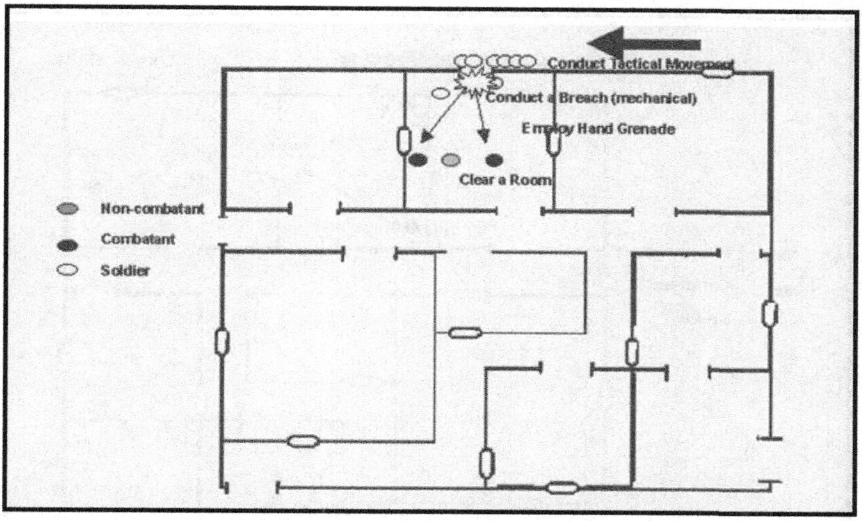

Figure 3-3. Event 1, *Squad Scenario (Crawl)*, force-on-targetry.

Table 3-3. Event 2, Squad Sequence (*Walk*).

Element	*Event Action*	*Estimated Time*
2 (WALK)	**Terminal Learning Objective** Familiarize the squad with how to enter and clear multiple rooms during the walk phase of training **Tasks** *Enter and Clear a Building (Squad)* • *Perform Movement Technique* • *Select Hasty Firing Position* *Conduct a Breach (Explosive DES)* • *Engage Targets with M16A2/M4* • *Engage Targets with M249 Machine Gun* *Conduct AAR*	**2 Hours**
NOTES: 1. Perform controlled pairs and target discrimination techniques (FM 3-22.9) **2. Include techniques (FM 3-06.11) and performance measures (FM 3-21.8 and DTMS).**		

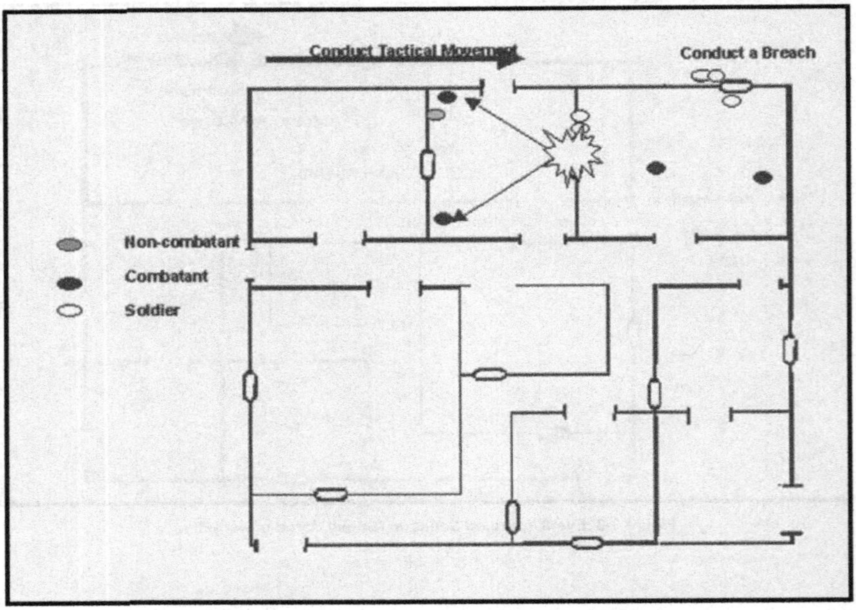

Figure 3-4. Event 2, *Squad Scenario (Walk)*, force-on-force.

Table 3-4. Event 3, Squad Sequence (*Run*).

Element	*Event Action*	*Estimated Time*
3 (RUN)	**Terminal Learning Objective** Refamiliarize squad with how to enter and clear multiple rooms during the run phase of training. **Tasks** *Enter and Clear a Building (Squad)* • *Perform Movement Technique* • *Select Hasty Firing Position* *Conduct a Breach (Mechanical)* • *Engage Targets with M16A2/M4* • *Engage Targets with M249 Machine Gun* *Clear a Hallway* • *Engage Targets with M16A2/M4* • *Engage Targets with M249 Machine Gun* *Conduct a Breach (explosive DES)* • *Engage Targets with M16A2/M4* • *Engage Targets with M249 Machine Gun* *Conduct AAR*	**1 Hour**

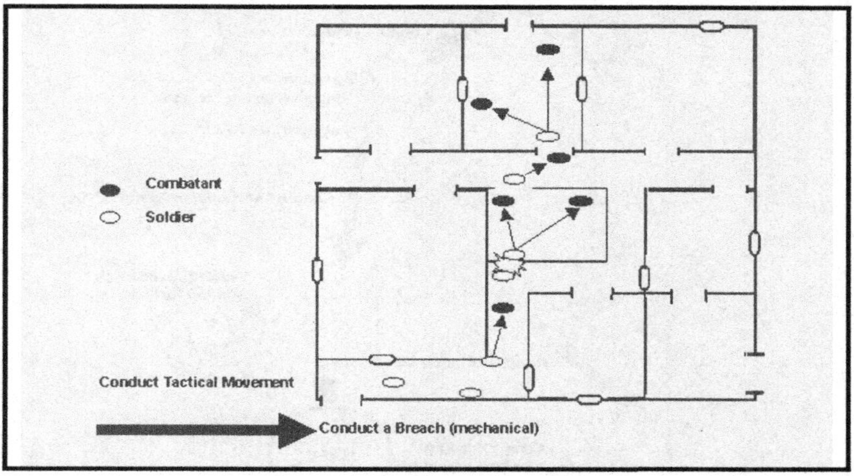

Figure 3-5. Event 3, *Squad Scenario (Run)*, force-on-targetry.

Table 3-5. Event 4, Platoon Evaluation Sequence (*Run*).

Element	*Event Action*	*Estimated Time*
4 (RUN)	*Occupy Assembly Area*	1 Hour
	Develop and Communicate a Plan	2 Hours
	Prepare for Combat	30 min
	Perform Passage of Line	1 Hour
	Conduct Tactical Movement	
	Break Contact	5 min
	Conduct Intermediate AAR	20 min
	Conduct Tactical Movement	
	Perform Overwatch/Support by Fire	1 Hours
	Assault a Building	1 Hours
	Consolidate and Reorganize	2 Hours
	Conduct Final AAR	1 Hours

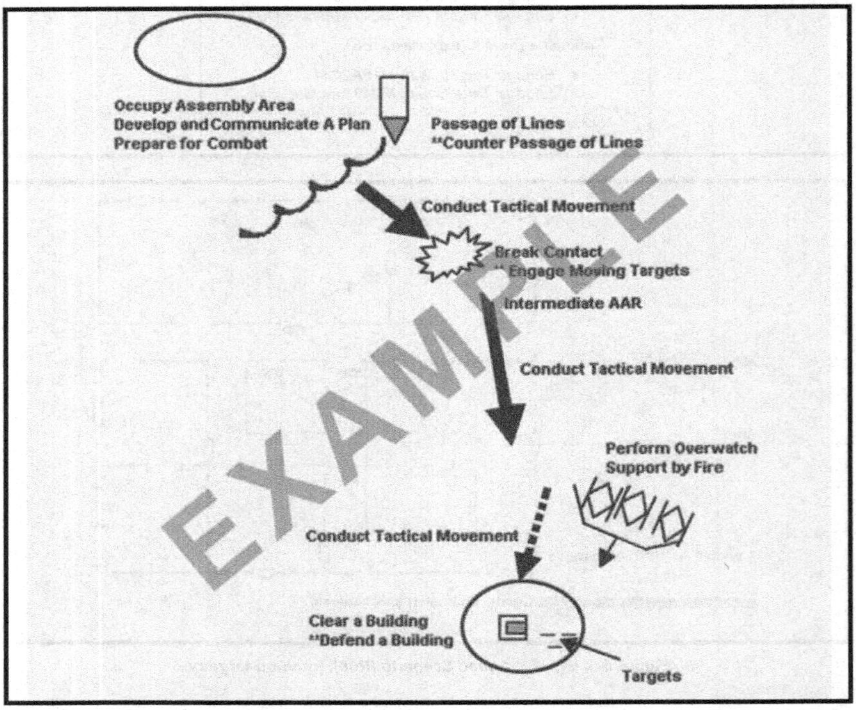

Figure 3-6. Example platoon tactical scenario (*Run*).

OPERATION ORDER

3-7. Figure 3-7 shows an example OPORD for the conduct of training.

```
                    ______________
                    (Classification)
                                        Copy ____ of ___ copies
                                       ____Company _____ Infantry
                                      (Unit Location)___________
                                       Date/time Group __________

(Message Reference Number)

OPORD: _____________
Reference Map, Series ___________________
Time Zone used Throughout the Order:_____________________

TASK ORGANIZATION:

______Platoon Platoon
______Platoon ____ 60-mm Section
______Antiarmor Section
```

1. SITUATION

a. **Enemy Forces**. Annex B (Intelligence Overlay [prepared by battalion S-2]).

(1) The company is opposed by elements of the Red Company of the Blue Brigade at about 60 percent strength. Within our zone, we have a platoon-sized element occupying a defensive position in a building in the vicinity of (______). The enemy has small arms weapons and may call on company mortars for support.

(2) Current indications are the enemy is regrouping to initiate small-unit offensive actions in the area. His most likely course of action is to move to the (direction) to link up with elements of the parent unit.

b. **Friendly Forces**.

(1) (___) Infantry attacks at (______) hours to seize Objective ALPHA (grid). The intent is to prevent the enemy element from moving (direction) to link up with other enemy units.

(2) Company (__) on our (left, right) makes the main attack at (____) hours to seize Objective RED (grid) in order to establish blocking positions to stop the enemy's movement to the (direction).

____
(Classification)

Figure 3-7. Example company OPORD.

(Classification)

2. MISSION

Company (__) attacks to seizes Objective BLUE (grid) at (date/time) hours in order to prevent the enemy from regrouping and engaging Company (__) (main effort) from Objective BLUE.

3. EXECUTION

a. **Concept of Operation**. Annex C (Operations Overlay). (This should include those steps listed in the task , Conduct Troop-Leading Procedures, paragraph 17. These should be written out when possible and not just placed on an overlay.)

(1) Maneuver. (___) Platoon establishes a defense in its zone, NLT (date/time) hours, to destroy small enemy elements moving or exfiltrating to the (direction), from the vicinity of Objective BLUE. This defense supports the attack on Objective BLUE by completing destruction of any enemy that may escape. (Evaluated) Platoon conducts the company main attack at (hours) to seize Objective BLUE to destroy the enemy or force them to withdraw into (___) Platoon's defense. (___) Platoon supports the attack of the (Evaluated) Platoon on Objective BLUE to suppress enemy fires and isolate the objective. On seizure of Objective BLUE, (Evaluated) Platoon and (___) Platoon will defend in sector to destroy the enemy elements withdrawing to avoid (___) Company's attack.

(2) **Fire support**.

(a) Priority of fires.

- Initially to (Evaluated) Platoon.
- Upon seizure of Objective BLUE, priority shifts to (Defensive) Platoon.

(b) Priority targets.

- 60-mm priority target initially AA4031.
- On initiation of assault on Objective BLUE, priority shifts to target AA4027 to isolate the objective.
- Upon seizure of Objective BLUE, priority shifts to target AA4025.

(Classification)

Figure 3-7. Example company OPORD (continued).

(Classification)

b. **Tasks to maneuver units.**

(1) (___) Platoon. Destroy enemy elements moving to the (direction), NLT (date/time).

(2) (Evaluated) Platoon. Consolidate and defend in sector to destroy enemy elements following the seizure of Objective BLUE.

(3) (___) Platoon. Be prepared to assault Objective BLUE, on order, consolidate and defend in sector to destroy enemy elements following the seizure of Objective BLUE.

c. **Tasks to Subordinate Units.**

(1) 60-mm Section. Initially in the vicinity of CP 7, priority of fires to the (Evaluated) Platoon. Move with the (Supporting) Platoon. On order, priority of fires to the (Defensive) Platoon.

(2) Antiarmor Section. Attached to the (Supporting) Platoon.

d. **Coordinating Instructions**:

(1) (Defensive) Platoon reports defense established and locations.
(2) Platoon reports the seizure of all objectives.
(3) Platoon reports the consolidation and establishment of the defense.
(4) Order of march from the assembly area to the objective: defense platoon, evaluated platoon, headquarters, support platoon, 60-mm mortar.

4. **SERVICE SUPPORT**

a. **General**. Company trains collocated with the battalion combat trains in the vicinity of (grid). Supply distribution at the company CP.

(Classification)

Figure 3-7. Example company OPORD (continued).

(Classification)

b. **Materiel and Services.**

(1) Class I: Ration cycle MRE-MRE-MRE.

(2) Class V: Pick up basic load at company CP, NLT (date/time).

(3) Medical: Company casualty collection point is __________.

(4) EPW: EPW collection point is ___________.

5. **COMMAND AND SIGNAL**

a. **Command.** Company command group initially follows the (Evaluated) Platoon.

b. **Signal.**

(1) Radio-listening silence is in effect until contact is made.

(2) SOI _____ is in effect.

JONES
Commanding

AnnexES:
B-Intelligence Overlay
C-Operations Overlay
D-Fire Support Overlay

(Classification)

Figure 3-7. Example company OPORD (continued).

TRAINING SUPPORT REQUIREMENTS

3-8. Minimum trainers and evaluators should be based on unit proficiency, training needs, and local safety requirements. Table 3-6 contains consolidated support requirements.

Table 3-6. Consolidated support requirements.

Ammunition	*Event 1*	*Event 2*	*Event 3*	*Event 4 (Optional)*
5.56-mm (per weapon) ball, blank, SRTA, or SESAMS	20 Blank	20 SESAMS	20 Ball/SRTA	20 Ball/SRTA
5.56-mm linked (per weapon)	30 Blank	30 SESAMS	30 Ball/SRTA	30 Ball/SRTA
Hand grenade body (practice)	0	--	4	4
Hand grenade fuzes (practice) per platoon	0	--	4	4
Hand grenade, smoke, per platoon	0	6	6	6
M84 hand grenade, stun, per platoon	0	4	4	0
TES (MILES Equipment)				
M16 (1 per weapon, 1 per OPFOR) M249 (1 per weapon) SAAF (1 per platoon) Controller Gun (1 per OC)	NO NO NO NO	YES YES YES YES	NO NO NO NO	NO NO NO NO
Targetry	5	0	5	10
OPFOR	NO	6	NO	NO

TRAINING AND EVALUATION

3-9. The T&EOs are the foundation of the MTP and the collective training of the platoon. T&EOs are training objectives (task, conditions, and standards) for the collective tasks that support platoon critical wartime operations. T&EOs may be trained separately, in an STX, in an FTX, or in live-fire exercises. For collective live-fire standards, the trainer needs to refer to the applicable Infantry Gunnery Manual for the appropriate course of fire. Those standards and courses of fire need to be integrated into the training exercise. Table 3-8 provides a list of related collective and supporting individual tasks.

Table 3-7. Related collective and supporting individual tasks.

Task	FM 3-21.8	FM 3-06.11	STP 21-1-SMCT	STP 7-11B1-SM-TB	STP 21-24-SMCT
Conduct Consolidation and Reorganization Treat Casualties Evacuate Casualties Process Enemy Prisoners of War/Captured Materiel Defend MOUT Building Conduct Tactical Movement	X	X			
Supporting Individual Tasks					
Engage Targets with an M16-Series Rifle			071-311-2007		
Engage Targets with an M16-Series Rifle Using a Night Vision Sight AN/PVS-4			071-315-2007		
Engage Targets with an M249 Machine Gun				071-010-0006	
Employ Hand Grenades			071-325-4407		
Move Over, Through, or Around Obstacles (Except Minefields)			071-326-0503		
Neutralize Booby Traps					052-193-1013
Move as a Member of a Fire Team				071-326-0501	
Perform Movement Techniques during MOUT		X		071-326-0541	
Conduct the Maneuver of a Squad				071-326-5611	
Conduct an Explosive Breach				071-440-0030	
Execute a Mechanical Breach				071-440-0027	
Conduct a Breach				071-440-0029	
Operate Night Vision Goggles AN/PVS-7				071-710-0008	
Select Hasty Firing Positions during MOUT				071-326-0557	
Control Movement of a Fire Team				071-326-5605	

Chapter 4

Breach Facility

Urban operations require Soldiers to enter buildings through walls, doors, windows, and roofs. The Breach Facility is designed to train Soldiers using tasks and techniques on how to breach locked doors, windows, and create man-sized holes in walls. The facility can be used to train mechanical, ballistic, thermal, and explosive breaching.

PURPOSE

4-1. The Breach Facility (Figure 4-1) is used to train Soldiers semiannually on the technical aspects of breaching techniques. It is also used to train TTP and explosive techniques not trained on any other type range.

Figure 4-1. Breach facility.

DESCRIPTION

4-2. This facility has three stations:

STATION 1

4-3. This station is a facade that is 8 feet high and 88 feet wide with four sections of two doors for a total of eight doors. Soldiers can conduct mechanical, thermal, ballistic, and explosive breaching techniques on doors at this station. Half of the doors open from the exterior, and the other half of the doors open from the interior (Figure 4-2).

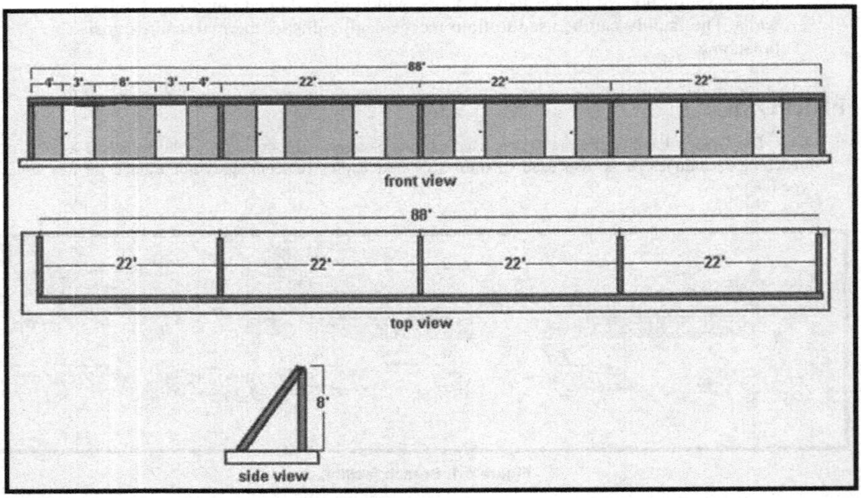

Figure 4-2. Station 1—Door breaching.

STATION 2

4-4. This station is a facade. It measures 10 feet high by 88 feet wide. It has four sections of two windows each, for a total of eight windows. Soldiers can conduct mechanical and explosive breaching techniques on windows at this station (Figure 4-3).

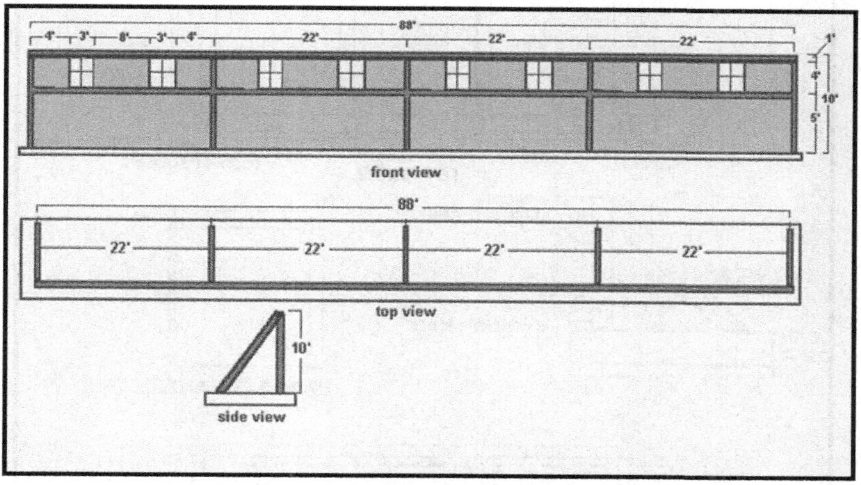

Figure 4-3. Station 2—Window breaching.

STATION 3

4-5. This station is a facade that looks like a wall of a building. It measures about 8 feet high by 28 feet wide. Precast panels will fit into each of the three sections. The panels measure 8 feet square and can be made of concrete, concrete blocks, stone, brick, wood or other material. The panels are designed for use with thermal and explosive breaching techniques (Figure 4-4).

TRAINING AUDIENCE

4-6. The facility provides Infantry, Engineer, MP, and Cavalry leaders with a facility to train and evaluate individual and leader breaching tasks. Unit training requirements determine the size of the unit and the tasks to be trained at this facility (Appendix B provides a unit training matrix). The individual tasks trained at the Breach Facility follow:

- 052-193-1003, *Prime Explosives Nonelectrically.*
- 071-440-0027, *Execute a Mechanical Breach.*
- 051-193-1055, *Construct a Nonelectric Initiating Assembly with Modernized Demolition Initiators (MDI).*
- 052-193-1312, *Construct Demolition Initiating Sets.*
- *Conduct a Thermal Breach* (Engineers only).

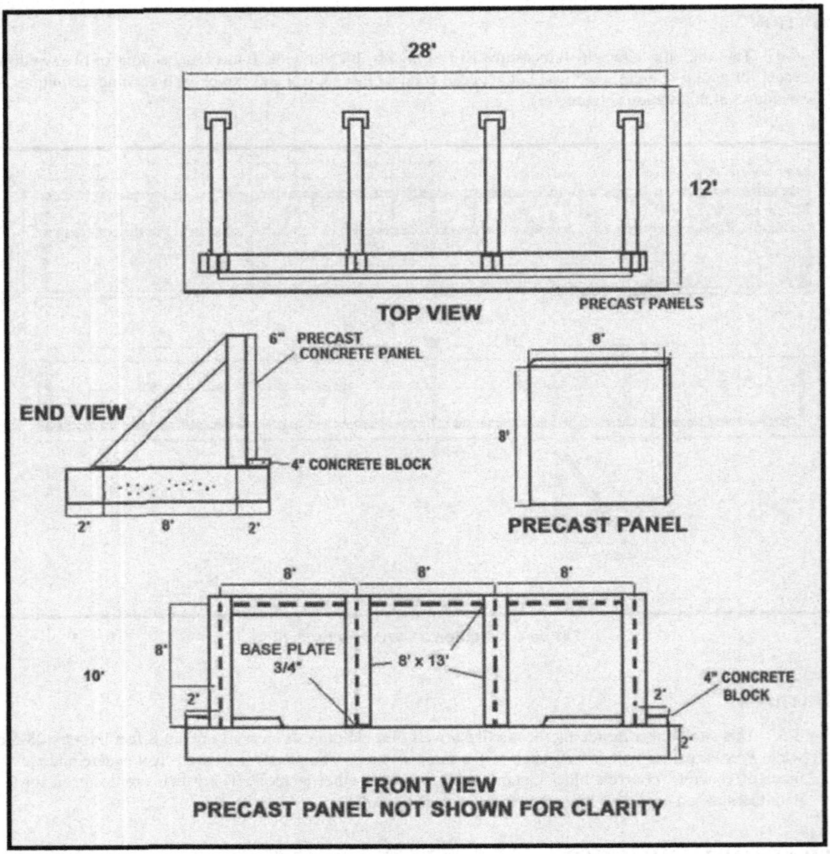

Figure 4-4. Station 3—Wall breaching.

TRAINING INFORMATION

4-7. Many types of organizations can train at this facility. Tactical operations in an urban environment may range from combat missions to stability and support actions. Commanders should note that this facility is an individual task and technique trainer. Training at the Breach Facility should be conducted prior to conducting collective training at the UAC, shoot house, and the CACTF. Trainers *must be certified* before conducting demolition training per local SOP or ACOM policy.

CONDUCT OF TRAINING

4-8. Trainers develop realistic time requirements for each task to be trained based on the *crawl-walk-run* concept and the size of the unit to be trained. The training event may be arranged in any order as long as it builds from simple to complex task and techniques. Table 4-1 contains an example sequence of events to assist units in developing their training plan. Leaders should do the following prior to conducting training. Before conducting ballistic breaching with automatic weapons, leaders should consult FM 3-06.11:

- Identify the unit's critical missions (see commander's training guidance) and select those tasks that support specific METL related to urban operations.
- Arrange the related tasks in the order of their importance to mission accomplishment.
- Assess the unit's current level of proficiency (strengths and weaknesses) and determine the degree of difficulty of each task. The training events should start from the less difficult tasks to the more difficult tasks. This instills confidence in the Soldiers conducting the training. Leaders should consult FM 3-06.11 for more information on explosive breach techniques.
- Identify the individual and leader tasks that are to be trained.
- Set up training conditions based on unit proficiency.

Table 4-1. Example platoon sequence of events.

Element	*Event Action*	*Estimated Time*
3d Sqd	Wall Breach: Explosive (crawl, walk, run)	3 hours
1st Sqd	Door Breach: Mechanical, ballistic, explosive (crawl, walk, run)	
2d Sqd	Window Breach: Mechanical, explosive (crawl, walk, run)	
2d Sqd	Wall Breach: Explosive (crawl, walk, run)	3 hours
3d Sqd	Door Breach: Mechanical, ballistic, explosive (crawl, walk, run)	
1st Sqd	Window Breach: Mechanical, explosive (crawl, walk, run)	
1st Sqd	Wall Breach: Explosive (crawl, walk, run)	3 hours NOTE: Walls, doors, and windows require replacement after each run/iteration to train to standards.
2d Sqd	Door Breach: Mechanical, ballistic, explosive (crawl, walk, run)	
3d Sqd	Window Breach: Mechanical, explosive (crawl, walk, run)	

TRAINING SUPPORT REQUIREMENTS

4-9. Table 4-2 shows minimum requirements based on three breach teams for each platoon-sized element. This facility is resource intensive, and Table 4-3 shows the total annual requirements per platoon-sized element.

Table 4-2. Consolidated minimum support requirements.

Ammunition/Equipment	*Crawl*	*Walk*	*Run*
Mechanical Breach: *Prying , Striking, or Cutting Tool*			
Door	(talk through)	1 per team (defeat one lock)	1 per team (defeat hinges)
Window	(talk through)	1 per team (defeat one lock)	1 per team (defeat hinges)
Ballistic Breach: Door			
Shotgun [1]	(talk through)	2 rds per weapon (defeat one lock)	5 rds per weapon (defeat hinges)
GREM (inert) [2]	(talk through)	1 per team – inert	1 per team - service
[1] If available to unit. [2] Grenade Rifle Entry Munition (GREM).			
Explosive Breach:			
Door: Det Cord (Flex Linear Charge) Nonelectric Firing System	(build)	(practice emplacement)	29 feet per team 1 per team
Wall: C4 (Satchel Charge) Nonelectric Firing System	(build)		4 blocks per team
		(practice emplacement)	23 feet per team 1 per team
Window: Det Cord (Flex Linear Charge) Nonelectric Firing System	(build)		29 feet per team 1 per team
		(practice emplacement)	
Thermal Breach Equipment Sappers (Only)	1	1	1

Table 4-3. Consolidated annual support requirements.

Ammunition/Equipment	*Total*
Doors	6
Windows	6
Wall Panels	6
Shotgun Rounds	42
GREM	6 inert/2 service
Det Cord (feet)	486
C4 (blocks)	24
Nonelectric Firing Systems	18

TRAINING AND EVALUATION

4-10. Individual training is the first step in the building-block approach and sets the basis for unit or collective training. Mastery of individual skills is an essential precondition for urban operations training and a good method for conducting individual training is through concurrent training. Concurrent training is designed to train groups of Soldiers on different tasks or techniques and these tasks or techniques may or may not be related. For example, a leader may subdivide the unit at the Breach Facility into breaching orders. One group of Soldiers may be conducting an explosive breach, while other groups are conducting mechanical and ballistic breaches. Ballistic breaching and explosive breaching does have a SDZ, an MSD, and ricochet considerations.

SAFETY CONSIDERATIONS

4-11. Safety requires effort on the part of everyone. It is not enough to teach safety rules; the rules must be practiced and enforced. Depending on the nature of the obstacle, the breacher may use mechanical, ballistic or explosive techniques. The following guidelines should be followed when training at the Breach Facility.

- Wear all appropriate personal protective equipment and follow appropriate guidelines:
  -- Safety glasses.
  -- Hearing protection.
  -- Body armor.
  -- Helmet.
  -- Gloves.
  -- Elbow and knee pads.
- Inspect equipment before use.
- Use appropriate tool.
- Use appropriate amount of explosive not to exceed four blocks of C4. Place the C4 on the wall only, not on the foundation, as this would cause irrepairable damage.
- Stay within the surface danger zone.
- Always use the minimum safe distance.
- Do not modify tool.
- Ensure proper use and maintenance of hydraulic and pneumatic tools; tools can burst under pressure when improperly used.

Chapter 5

Combined Arms Collective Training Facility

The home station CACTF is designed to conduct multiechelon, full-spectrum operations training up to battalion TF level. Example company and platoon STX lanes, and stability and support operations tasks, are provided to help units develop training plans. Leaders use training exercises to train, evaluate, and practice performance-oriented collective tasks. TEWT, STX, and FTX are the three types of exercises covered in this chapter. Each is designed to help the trainer develop urban operations training skills and increase unit proficiency in an urban environment. The urban operations collective training exercises were developed using the standard CACTF as described in this chapter and may be modified to fit available urban operations facilities.

PURPOSE

5-1. The leader or commander is provided with a CACTF to evaluate unit urban operations proficiency. The CACTF is intended to support blank fire, MILES/TES, SESAMS, STX, and FTX scenarios on a semiannual basis. The CACTF supports branch-specific lane training and combined arms training up to battalion level across the full spectrum of operations: offense, defense, stability, and support.

DESCRIPTION

5-2. The CACTF replicates an urban environment. The facility consists of 2.25 square kilometers of urban sprawl with 20 to 26 buildings, roads, alleys, parking areas, underground sewers, parks, athletic fields, and command and control building. The actual size and configuration of the CACTF depends on the local installation site requirements. The CACTF is designed to support heavy and light Infantry, Armor, artillery, and aviation positioning and maneuver. Table 5-1 shows the recommended types of buildings and features in the CACTF.

Table 5-1. CACTF characteristics.

20 to 26 Buildings (1.5 km by 1.5 km)	
Tunnel and sewer system Shantytown One 3-story building Three 2-story buildings Industrial area Electricity and potable water City dump	Props and furniture Targetry Audio and image capture with EDIT/REPLAY capability Control building with AAR facility Breachable walls

5-3. The buildings have one, two, or three stories (some with basements), sloped and flat roofs. CACTF building variations pose different tactical and technical training challenges. Figure 5-1 shows an example CACTF containing 24 buildings with the following characteristics:

- The hotel is a dominating structure typical of a central business district. The hotel has an elevator shaft, fire escapes, and a large first-floor lobby.
- The two-story office building contains a series of identical rooms.
- The school has a long, central corridor, large windows, and a single large room.
- The townhouse has multiple sections with a common attic.
- The service station accommodates combat vehicles in its service bay.

5-4. The buildings have other features that enhance the CACTF's training value.

- Constructed mouse-holes permit movement between rooms and floors.
- Loopholes in roofs and outside walls allow observation and fields of fire.
- Roof hatchways lead to the top floors of multiple story buildings.

5-5. Rooms vary in size, and in number of doorways and windows. These variations require Soldiers to identify rooms that provide suitable indoor fighting positions for TOW, Javelin, Dragon, and AT4 antitank weapon systems.

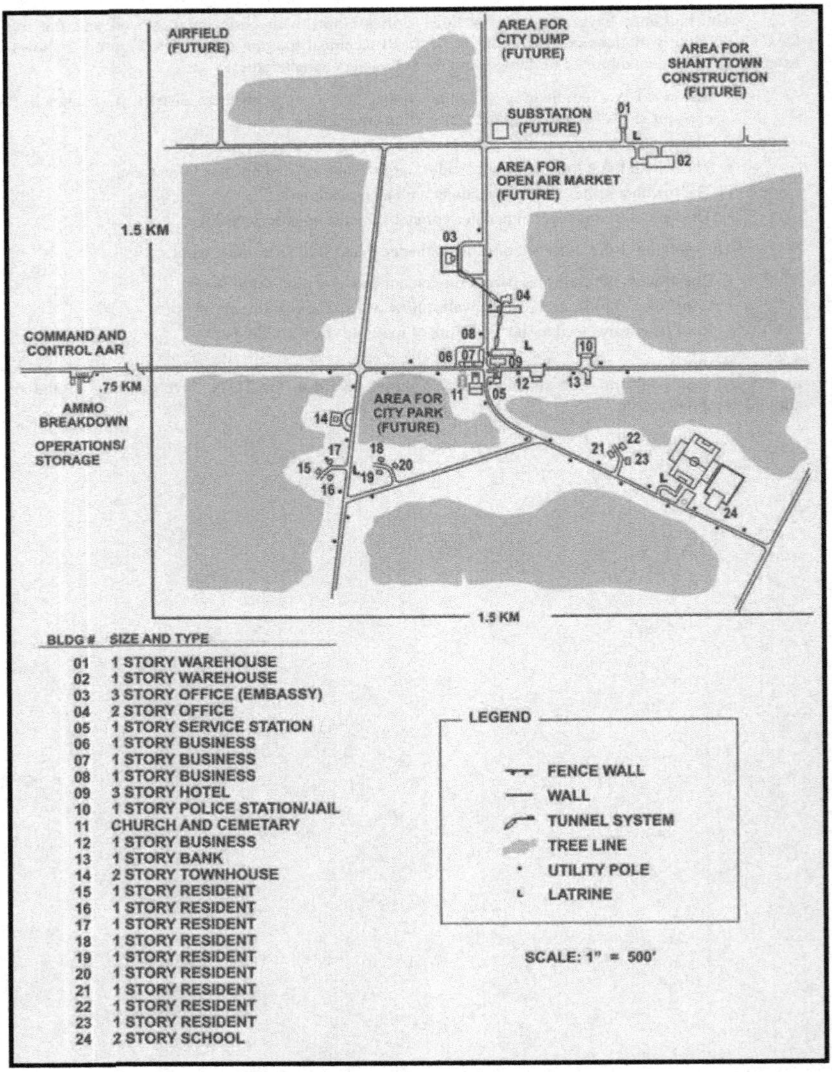

Figure 5-1. Example CACTF schematic.

5-6. Buildings are arranged in a realistic urban pattern. The CACTF represents an area consisting mainly of residential, commercial, public institutional, and light industrial buildings. Buildings and streets are arrayed with sufficient dispersion to allow up to nine platoon-sized STX sectors (Figure 5-2) or six company/team STX sectors (Figure 5-3).

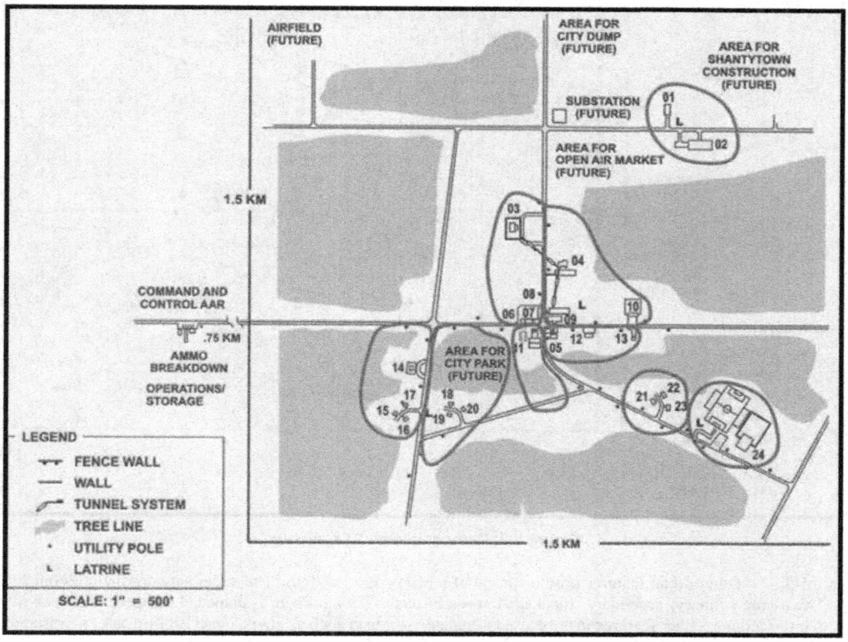

Figure 5-2. Platoon ODSS STX sectors.

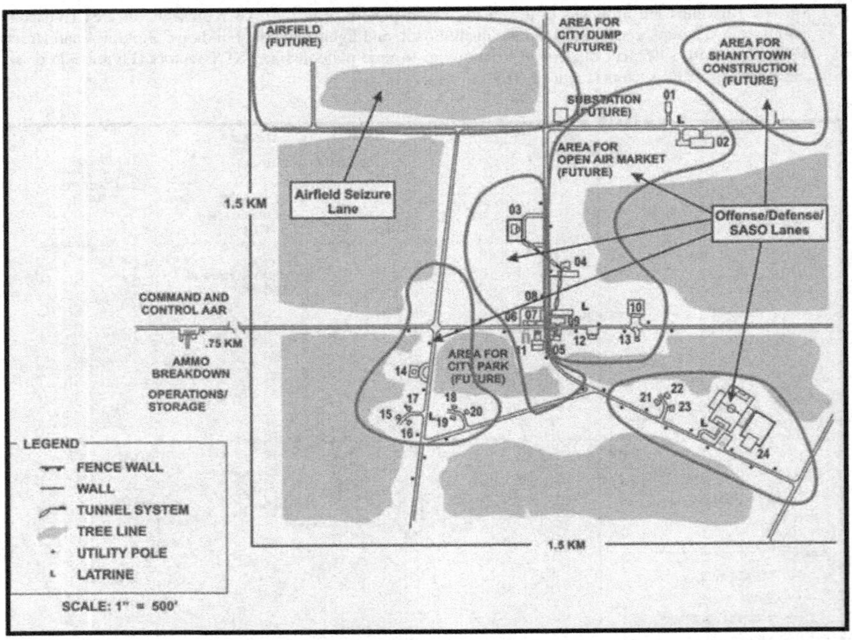

Figure 5-3. Company/team STX sectors.

5-7. Other urban features pose a variety of terrain considerations. The street network includes one and two-lane primary, secondary, local, and service roads. They join in T-shaped, L-shaped, and four-way intersections. Other features may be used to enhance realism such as signs, benches, furniture, newsstands, utility poles, street lights, vegetation, simulated rubble, live civilians on the battlefield, role players, and OPFOR. Open areas next to the CACTF can be used for maneuvering and staging areas. The specific details and layout of the CACTF may vary from one installation to another.

CAUTION

Trainers must ensure that fighting position reinforcement does not exceed the weight limits of the buildings in order to avoid structural damage and safety hazards to the Soldiers.

TACTICAL EXERCISE WITHOUT TROOPS

5-8. A terrain walk and a TEWT may be conducted in a local downtown area to prepare leaders and to take advantage of the available time in the CACTF. Terrain walks and TEWTs are conducted to teach leaders the tactical value of urban terrain. Appendix A provides training tips and additional information regarding the conduct of a TEWT.

COLLECTIVE TRAINING EXERCISES

5-9. The following exercises were developed IAW the guidance in FM 7-0, FM 7-1, FM 25-4, and TC 7-98-1. These publications include offensive and defensive scenarios for combat in urban areas, specifically, two platoon STXs and two company FTXs. Each can be conducted with blank ammunition; with MILES/TES; with 9-mm or 5.56-mm SESAMS for short-range precision target engagements (less than 50 meters); or force on force. Each was chosen to help the commander develop, sustain, and evaluate his unit's mission proficiency:

- Infantry Rifle Platoon Attack STX
- Mechanized Infantry Platoon Defend STX
- Infantry Rifle Company Attack FTX
- Mechanized Infantry Company Team Defend FTX

INFANTRY RIFLE PLATOON STX ATTACK

1. **Objective.**

This example STX trains collective, leader, and individual tasks for the platoon operation, Attack in a Built-Up Area.

2. **Interface.**

This STX is supported by two warrior drills, *React to Contact* and *Enter and Clear a Building*.

3. **Training.**

a. *Guidance*. FM 7-0 and FM 7-1 provide training guidance, tips, and enhancers for planning and executing STXs. The trainer should review the individual, leader, and collective tasks to be performed during the STX to determine the ones that require initial or refresher training.

b. *General Situation.*

(1) The exercise scenario follows: Contact with the enemy has been reestablished. Initial reports indicate he is at 65 to 70 percent strength, and that he has not received reinforcements. His defensive positions are in the CACTF and are not well established. He has indirect fire and CAS; he has already used chemicals, and will probably do so again. The platoon receives an order to prevent the enemy from establishing a heavily fortified defense in the CACTF. Such a defense could prevent friendly forces from continuing offensive operations.

(2) This exercise begins when the platoon leader receives the company FRAGO, and it ends after the platoon consolidates and reorganizes on the objective. An AAR should be held after completing the task, *React to Contact*, and at the completion of the exercise. If necessary, portions of the exercise should be repeated until the platoon performs them to standard. Figure 5-4 provides a graphic scenario of task performance for the STX *Attack*. Table 5-2 provides a recommended sequence for performance of T&EOs, and the time allotted for each portion of the STX.

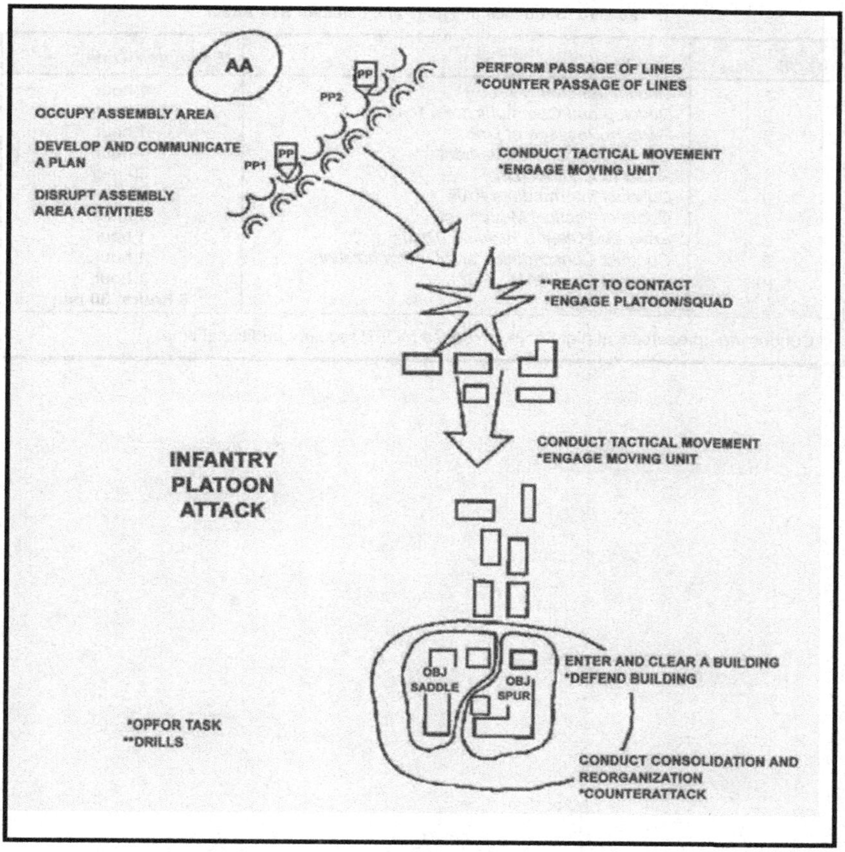

Figure 5-4. Attack STX scenario.

Table 5-2. Sequence of events and times for STX Attack.

Event	*Action*	*Estimated Time*
1	*Occupy Assembly Area*	1 hour
2	*Develop and Communicate a Plan*	1 hour
3	*Perform Passage of Lines*	1 hour
4	*Conduct Tactical Movement*	1 hour
5	*React to Contact (Drill)*	30 min
6	*Conduct Intermediate AAR*	30 min
7	*Conduct Tactical Movement*	30 min
8	*Enter and Clear a Building (Drill)*	1 hour
9	*Conduct Consolidation and Reorganization*	1 hour
10	*Conduct Final AAR*	1 hour
	*** Total Time**	**8 hours, 30 min**
* Conducting an exercise at night or in increased MOPP requires additional time.		

c. *Special Situation.* The platoon is part of a company in a secure assembly area when a FRAGO to attack is received (Figure 5-5).

1. **SITUATION.** The company is opposed by one platoon.

a. *Enemy Forces.* The enemy is at 65 to 70 percent strength. He is preparing a defense in CACTF (**MOUT site**) to prevent friendly forces from continuing to attack along the CACTF avenue of approach. He is expected to use artillery-delivered nonpersistent nerve agents in the defense of CACTF.

b. *Friendly Forces.* (Battalion designation) attacks (**date-time group**) to destroy enemy forces at OBJ COWBOY (**CACTF**) in order to pass follow-on forces, allowing friendly forces to continue the attack.

2. **MISSION.** (___) Company seizes OBJ HORSE (CACTF) vicinity (**grid**) NLT (**date-time group**) to prevent the enemy from establishing a heavily fortified defense.

3. **EXECUTION.**

a. *Concept of the Operation.* (See overlay.). The company moves in a wedge to Assault Position 1, assault CACTF to seize OBJ HORSE, consolidate, and reorganize.

(1) Maneuver. (**Evaluated**) Platoon will lead the company and be the main effort in the attack. They seize OBJ SADDLE vicinity (**grid**) to allow passage of friendly elements through CACTF. (___) Platoon moves on the company's left and seize OBJ SPUR vicinity (**grid**) to protect the left flank of the main effort. (___) Platoon moves on the company's right and establish a support-by-fire position located vicinity (**grid**).

(2) Fire support. fires are used to obscure movement to OBJ HORSE and to suppress the enemy during the assault. Priority of indirect fires is to (*evaluated*) Platoon, which is allocated one priority target (**120-mm mortar**).

Figure 5-5. Example FRAGO for STX Attack.

```
 b. Tasks to Maneuver Unit.
    (1)  (Evaluated) Platoon.
       (a)  Use Passage Point 1.
       (b)  Orient defense from TRP 2 to TRP 3.
       (c)  Be prepared to employ a wire-mined obstacle vicinity TRP 5
during consolidation.
    (2)  (___) Platoon.
       (a)  Use Passage Point 2.
       (b)  Orient defense from TRP 1 to TRP 2.
    (3)  (___) Platoon
       (a)  Use Passage Point 2.
       (b)  Orient defense from TRP 3 to TRP 4.
 c. Coordinating Instructions.
    (1)  Company RP is located (grid).
    (2)  Company linkup is (grid).
    (3)  MOPP1 is in effect (date and time).
    (4)  Selective rubbling is authorized. Destruction of an area
larger than one city block requires brigade commander approval.
```

Figure 5-5. Example FRAGO for STX Attack (continued).

4. **Support Requirements.**

a. *Minimum Trainers/Evaluators*: This exercise can be conducted with the company commander or platoon leader acting as the trainer or primary evaluator. At least one more controller or evaluator is required with the OPFOR. Other platoons being trained or evaluated will increase support personnel requirements.

b. *Vehicles/Communications*: All vehicles organic to the platoon should be included. The OPFOR should also have some combat vehicles.

c. *Opposing Force*:. The OPFOR should consist of at least a reinforced squad.

d. *Maneuver Area*: A training area at least 4 by 15 km is needed for cross-country movement and infiltration, and the CACTF is the ideal objective area. The terrain should offer multiple covered and concealed approaches to the objective.

5. **Evaluation.**

Table 5-3 lists the T&EOs from DTMS used to evaluate this STX.

Table 5-3. T&EOs used to evaluate STX Attack.

Task/Function	*Task Number*
Perform Passage of Lines	7-3-1125
Conduct Tactical Movement	7-2-1342
Occupy Assembly Area	7-3-1136
Develop and Communicate a Plan	7-3-1605
Conduct Consolidation and Reorganization (Company/Platoon)	7-2-5027

MECHANIZED INFANTRY PLATOON STX DEFEND

1. **Objective.** This example STX trains collective, leader, and individual tasks for the platoon operation, *Defend Urban Area/Building.*

2. **Interface.** This STX is supported by two warrior drills: *React to Contact* and *Enter and Clear a Building.*

3. **Training.**

a. *Guidance.* FM 7-0 and FM 7-1 provide training guidance, tips, and enhancers for planning and executing STXs. The trainer should review the individual, leader, and collective tasks to be performed during the STX to determine which require initial or refresher training.

b. *General Situation.*

(1) The exercise scenario is as follows: Latest INTSUMs indicate that enemy forces are expected to use a high-speed avenue of approach to counterattack. The approach is dominated by a small urban area containing strongly constructed buildings. The enemy has the capability for indirect fire, CAS, and chemicals. A defense in and around the town is necessary to deny the enemy the use of the avenue of approach. The platoon, as part of a larger force, is ordered to occupy and prepare defensive positions that overlook the avenue of approach. Indirect fire is available.

(2) Conduct of the exercise begins when the platoon receives an OPORD to move to establish a platoon defensive position. An AAR should be conducted after Soldiers enter a building and clear a room, and a final AAR should be conducted once all evaluation notes are compiled. If necessary, portions of the exercise should be repeated until the platoon's performance is satisfactory.

(3) When the platoon receives a warning order to prepare to move to establish a defense, the platoon is part of a company, which is part of a task force, and is located in an assembly area. Figure 5-6 provides a graphic scenario of task performance for the Defend STX. Table 5-4 provides a recommended sequence for performance of T&EOs and the time allotted for each portion of the STX.

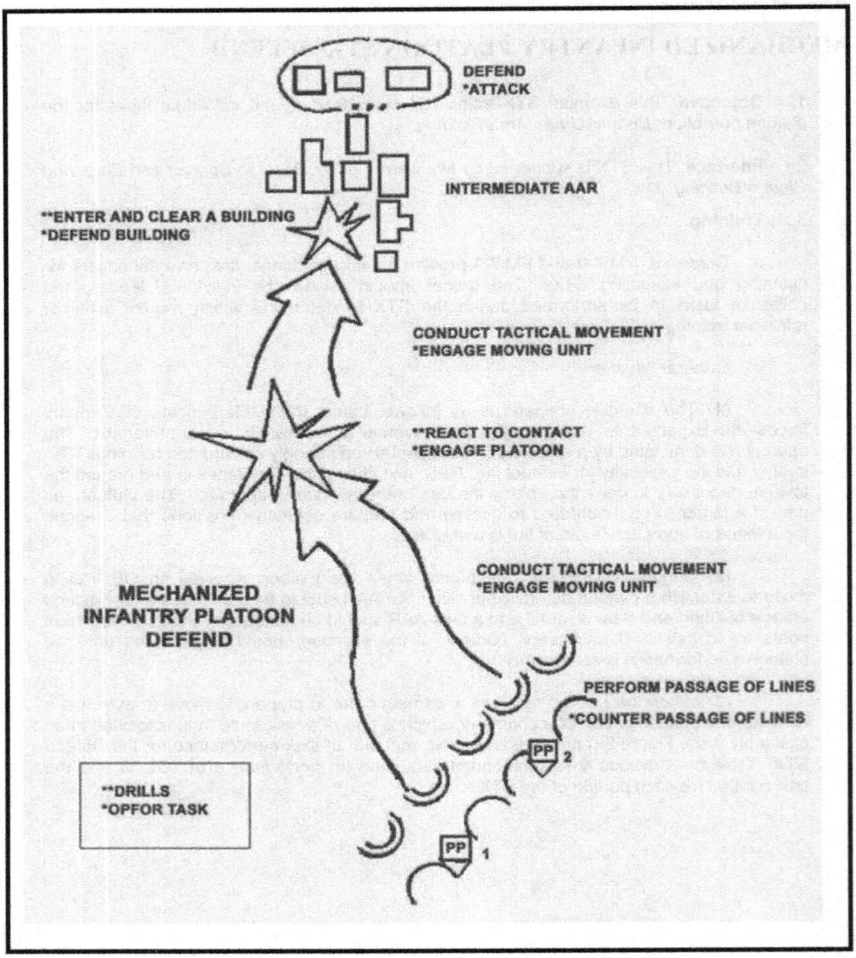

Figure 5-6. Defend STX scenario.

Table 5-4. Sequence of events and times for STX Defend.

Event	*Action*	*Estimated Time*
1	*Perform Passage of Lines*	1 hour
2	*Conduct Tactical Movement*	30 min
3	*React to Contact*	1 hours
4	*Conduct Tactical Movement*	30 min
5	*Enter and Clear a Building (Drill)*	1 hour
6	*Intermediate AAR*	30 min
7	*Defend MOUT/Building*	6 hours
8	*Conduct Final AAR*	1 hour
	***Total Time**	**11 hr 30 min**
*Conducting an exercise at night or in increased MOPP requires additional time.		

c. *Special Situation.* Thirty minutes after receiving the warning order, the platoon receives the company OPORD (Figure 5-7).

(Classification)

Copy no ___ of ___ copies
___ Battalion, ___ Infantry
Unit location or place of issue
Date-time group (with time zone)
Message reference number

OPERATION ORDER NO. _____
References: Map, series no. ___, sheet no. ___, edition ___
Time Zone Used Throughout the Order: ____
TASK ORGANIZATION:
___ Platoon, ___ Platoon(-), ___ Platoon, Company Control

Company HQ
1/1/C-52d Engr
1/2 ___,___
(Reserve)
Company Trains

1. **SITUATION.**

a. **Enemy Forces.**

(1) *Composition, Distribution, and Strength.* We are opposed by elements of the Liberation Guards Militia, Red Brigade. Several platoon-sized outposts are located within the battalion sector and consist of about 30 to 35 Soldiers per platoon. They are equipped with an assortment of Soviet small arms, including AK-47s, RPKs, RPGs, and various wheeled and tracked vehicles including some BMPs. Enemy outposts are at 60 to 70 percent strength and have indirect fire assets available.

(2) *Capabilities.* The enemy units are well entrenched in their defensive positions. The strongpoints have dedicated fire support assets and are capable of using nonpersistent nerve agents to prevent coordinated attacks. No enemy air threat exists but some heavy armored vehicles have been sighted in the area.

(3) *Probable Course of Action.* Intelligence indicates the enemy strongpoints will continue to fortify their positions until relief arrives within 24 hours in the form of a main enemy counterattack.

b. **Friendly Forces.** (__) TF defends in sector NLT (______) along PL (_____) to deny enemy forces use of avenue of approach (_____). (__) TF must be prepared to continue movement forward of PL (_____), on order.

(1) Missions of units on left and right are (as required).

(Classification)

Figure 5-7. Example OPORD for STX Defend.

(Classification)

(1) (__) TF defends in sector NLT (______) from (grid) to (grid) to (grid) to contain any penetrations of the brigade sector.

(1) (___) Battery, (___) Battalion provides DS to 1st Brigade.

c. **Attachments and Detachments**. 1/1/C-52d Engineers attached.

2. **MISSION**. (___) Company blocks enemy forces NLT (date-time group) to prevent them from using high-speed avenues of approach through (name of MOUT site), and to prevent the enemy from counterattacking into the battalion sector.

3. **EXECUTION**.

a. **Concept of the Operation**. (See overlay.) Company prevents the enemy from using the high-speed avenues of approach through (MOUT site) by moving to the site and establishing a blocking position, which overwatches the platoon BPs covering the main avenues. We will block in depth by placing two platoons forward and one back, and maintaining a reserve to block any penetration. We must ensure our sectors of fire are placed to mass fire at the decisive point.

(1) *Maneuver*. (Evaluated) Platoon acts as the main effort, blocking from BP (__), vicinity (grid), to prevent enemy penetration of PL (_____). (___) Platoon occupies BP (__), vicinity (grid), to block enemy movement into the main effort's left flank. (___) Platoon occupies BP (__) to prevent the enemy from making a coordinated attack on the main effort.

(2) *Fire Support*. The purpose of fires is to disrupt the enemy when his counterattack reaches PL (____). Priority of indirect fires is to (*evaluated*) Platoon, which is allocated two 155-mm priority targets, one of which may be used as an FPF.

(3) *Intelligence*. Priority of collection is to identify and locate enemy combat reconnaissance patrols and the enemy main effort.

(4) *Engineering*. During movement to the battle position, obstacles are used to ensure mobility. During blocking actions, the purpose of obstacles is to prevent the enemy from advancing on larger streets and to canalize him into more restrictive terrain. Priority of effort is to countermobility. Priority of support is to (*evaluated*) Platoon.

b. **Tasks to Maneuver Units**:

(1) *(Evaluated) Platoon*:

(a) Conduct passage of lines using PP 1.

(b) Lead the company movement after crossing the LD. Be prepared to react to enemy contact when (MOUT site) is reached.

(c) Enter and clear building(s) (number determined by commander).

(Classification)

Figure 5-7. Example OPORD for STX Defend (continued).

(Classification)

(d) Occupy BP (__) and orient on TRPs 2 and 3.
(e) Establish an OP vicinity (grid) to observe main avenue of approach.
(2) *(___) Platoon*:
(a) Conduct passage of lines using PP 2.
(b) Move second in order of movement.
(c) Enter and clear building(s) (number determined by commander).
(d) Occupy BP (__) and orient on TRP 1.
(3) *(___) Platoon*:
(a) Conduct passage of lines using PP 3.
(b) Move last in order of movement.
(c) Enter and clear building(s) (number determined by commander).
(d) Occupy BP (__) and orient on TRPs 4 and 5.
(e) Establish an OP vicinity (grid).

c. **Tasks to Subordinate Units**: Engineer squad travels with the company headquarters behind (*evaluated*) Platoon and supports the main effort's emplacement of obstacles.

d. **Coordinating Instructions**.
(1) Timings.
(a) Brief back ________.
(b) Company rehearsal ________.
(c) SP time ________.
(d) In position NLT (time) ________.
(2) Report all enemy contact.
(3) Report crossing all phase lines.
(4) MOPP status ___; enemy chemical attack is expected.
(5) Air defense status is YELLOW, WEAPONS HOLD.
(6) Leaders briefback the movement and defense plan to the commander at the company CP at (date-time group).
(7) Destruction of subterranean approaches requires brigade commander's approval.
(8) Selective rubbling is authorized. Destruction of more than one building requires commander's approval.
(9) Marking of rooms and buildings is IAW CO SOP.

(Classification)

Figure 5-7. Example OPORD for STX Defend (continued).

(Classification)

(10) Rules of Engagement follow:
(a) Minimize collateral damage to structures.
(b) Civilians have been evacuated. Evacuate to company CP any other civilians encountered.

4. SERVICE SUPPORT. IAW the battalion SOP.

a. **Miscellaneous.**
(1) Battalion trains located vicinity (grid).
(2) Company trains located vicinity (grid).

b. **Materiel and Services.**
(1) *Supply.*
(a) Class I. Ration Cycle M-M-M. Platoons must maintain a three-day supply in each BP.
(b) Class II. Submit requirements for special equipment ASAP. Coordinate with 1SG.
(c) Class III. Refuel all vehicles before (date-time group).
(d) Class IV. Coordinate with S-4 for delivery.
(e) Class V.
- Platoons cache one additional basic load in BP.
- Platoons submit request for additional demolitions NLT (date-time group).

(2) *Services.* Battalion decontamination site is located vicinity (grid).
(3) *Maintenance.*
(a) Priority to (*evaluated*) Platoon M2 BFVs.
(b) UMCP located vicinity (grid).

c. **Medical Evacuation and Hospitalization.**
(1) Battalion aid station located vicinity (grid).
(2) Company CCP located with company CP.

d. **Personnel.**
(1) No replacements for next 72 to 96 hours.
(2) Brigade mortuary affairs point located in BSA.
(3) EPW collection sites located at (grid) for battalion and (grid) for company.

(Classification)

Figure 5-7. Example OPORD for STX Defend (continued).

```
                         (Classification)

5.  COMMAND AND SIGNAL.

    a.  Command.  Initially, company CP will be traveling center of
company wedge. Location in BP to be determined.

    b.  Signal.
        (1) Current SOI in effect.
        (2) Primary means of communication in BP is wire, FM, messenger,
in that order.
        (3) Radio listening silence is in effect until after crossing PL
(____) or until contact is made.

                         (Classification)
```

Figure 5-7. Example OPORD for STX Defend (continued).

4. **Support Requirements**.

a. *Minimum Trainers/Evaluators*: This exercise is conducted by one company grade officer who is the trainer and primary evaluator.

b. *Vehicles/Communications*: All vehicles organic to the platoon should be included.

c. *Opposing Force*:. The OPFOR should be platoon strength.

d. *Maneuver Area*: A training area at least 2 by 10 km, with hasty fighting positions, is desirable for cross-country movement. The CACTF serves as the objective area. The terrain, ideally, should offer multiple covered and concealed routes. Using terrain that limits the leader to a geographic or predictable route prevents evaluation of the unit's ability to conduct a terrain analysis and select covered and concealed routes.

e. *Barrier Materials.* Sufficient Class IV (sandbags, lumber, and so on) must be made available to fortify positions, IAW Chapter 3, FM 3-06.11.

5. **Evaluation**.
Table 5-5 lists the T&EOs from DTMS used to evaluate this STX, and Table 5-6 lists the supporting tasks.

Table 5-5. T&EOs used to evaluate STX Defend.

Task/Function	*Task Number*
Defend MOUT/Building	7-3-1118
Perform Passage of Lines	7-3-1125
Conduct Tactical Movement (Co/Plt)	7-2-1342

INFANTRY RIFLE COMPANY FTX ATTACK

Table 5-6. Supporting tasks for company FTX.

STX C-II-1	*STX C-II-2*	*STX C-II-3*
Conduct an Infiltration (Co/Plt) Perform Linkup Execute an Assault (MOUT) Conduct Consolidation and Reorganization (Co/Plt)	Perform Cordon and Search Take Action on Contact Conduct an Infiltration (Co/Plt) Perform Linkup Perform Passage of Lines Occupy Assembly Area	Disarm Belligerents Handle Captured Insurgents or Belligerents Apprehend/Detain Noncombatants

1. **Objective**. This FTX is designed to train the Infantry rifle company in conducting offensive operations. This exercise provides practice for the company commander, platoon leaders, squad leaders, and company headquarters personnel in planning, coordinating, and controlling combat operations.

2. **Interface**. This FTX is supported by STX C-II-1, C-II-2, and C-II-3 (Table 5-6). It is also supported by two warrior drills, *React to Contact* and *Break Contact*.

3. **Training**.

a. *Guidance*. FM 7-0 and FM 7-1 provide training guidance, tips, and enhancers for planning and executing STXs. The trainer should review the individual, leader, and collective tasks to be performed during the STX to determine which require initial or refresher training.

b. *General Situation*.

(1) The battalion is ordered to conduct offensive operations in support of the brigade mission. The company is ordered to infiltrate enemy lines and to destroy an insurgent operations base located in the CACTF. The company must be prepared to remain in the CACTF to conduct cordon and search missions.

(2) This exercise should be conducted under various conditions.

(3) When the company receives the battalion warning order the planning process for the attack mission begins. The exercise ends when the company completes a passage of lines, occupies an assembly area, and completes preparations to continue its mission. Figure 5-8 shows a graphic scenario of task performance for the FTX Attack. Table 5-7 provides a recommended sequence for performance of T&EOs and the time allotted for each portion of the STX, including AARs.

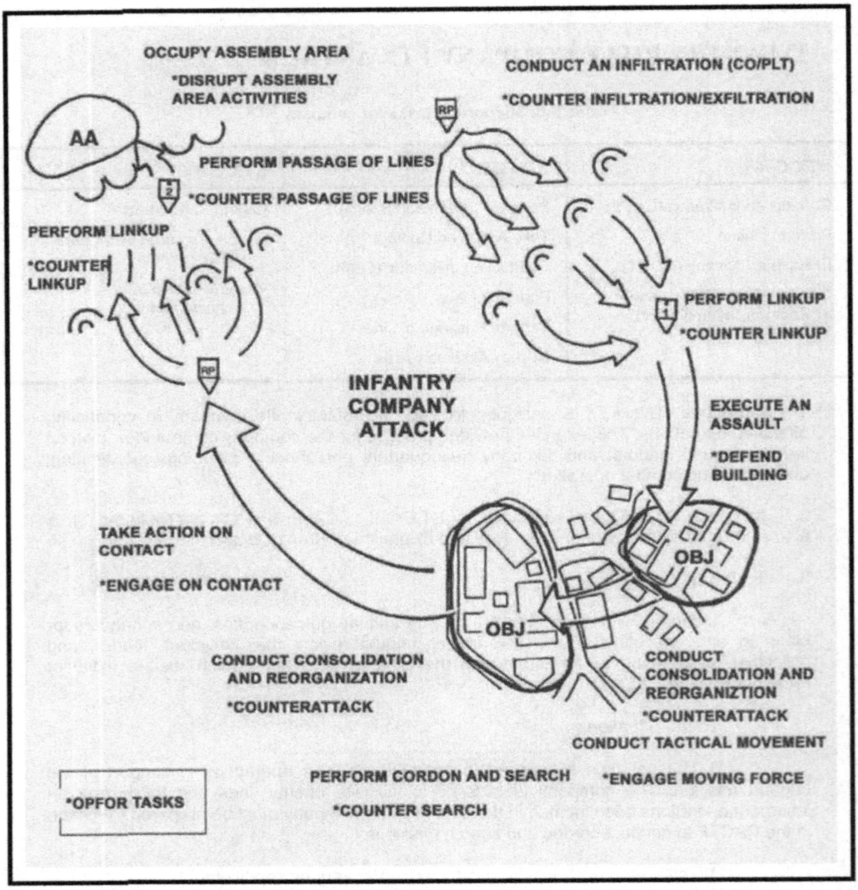

Figure 5-8. Attack FTX scenario.

Table 5-7. Sequence of events and times for FTX Attack.

Event	*Action*	*Estimated Time*
START STX C-II-1		
1	*Issue Company OPORD*	4 hours
2	*Conduct an Infiltration (Co/Plt)*	6 hours
3	*Perform Linkup*	3 hours
4	*Execute an Assault (MOUT) AAR*	3 hours
5	*Conduct Consolidation and Reorganization (Co/Plt)*	3 hours
ISSUE FRAGO, START STX C-II-2		
6	*Perform Cordon and Search*	6 hours
7	*Take Action on Contact / AAR*	3 hours
8	*Conduct an Infiltration (Co/Plt)*	6 hours
9	*Perform Linkup*	3 hours
10	*Perform Passage of Lines*	2 hours
11	*Occupy Assembly Area*	3 hours
START STX C-II-3		
12	*Disarm Belligerents*	3 hours
13	*Handle Captured Insurgents or Belligerents*	4 hours
14	*Apprehend/Detain Noncombatants*	4 hours
15	*Conduct Final AAR*	2 hours
	Total Time	**55 hours**

The time required to train an event varies based on the factors of METT-T and on the unit's training proficiency. Regardless, training is conducted to standards, not to time limitations. The following tasks are integrated and evaluated throughout the exercise:

- *Prepare for combat*
- *Maintain operations security*
- *Conduct tactical movement*
- *Perform consolidation*
- *Perform reorganization*
- *Perform stability and support operations*

* Time for intermediate AARs is excluded from this total; additional time is required if much of the exercise is conducted at night or during other limited visibility.

(4) The battalion commander issues an OPORD to the company commanders (Figure 5-9).

____
(Classification)

Copy no ___ of ___ copies
___ Battalion, ___ Infantry
Unit location or place of issue
Date-time group (with time zone)
Message reference number

OPERATION ORDER NO. _____
References: Map, series no. ___, sheet no. ___, edition ___
Time Zone Used Throughout the Order: ____

TASK ORGANIZATION:
___Company ___Company ___Company Battalion Control
Scout Platoon
81-mm Mortar Platoon
Antitank Platoon
1/C/58th ENGR (DS)
Battalion Trains

1. **SITUATION.**

a. **Enemy Forces**. Annex A (See intelligence overlay [prepared by brigade S-2].)

(1) *Disposition, Composition, and Strength.* The battalion is opposed by one company from the Dona Anna Militia Guards Battalion. This enemy company consists of 3 platoons with 35 to 40 Soldiers, equipped with M16s, AK-47s, and LMGs. The company headquarters is currently located in the town hall of CACTF. The company is at 90 percent strength and has indirect fire support assets available.

(2) *Capabilities.* The enemy possesses nonpersistent nerve agents and indirect fire assets capable of delivery. He also has some light armored vehicles in the area, but their location is undetermined. No enemy air threat exists. However, the enemy may be able to reinforce CACTF with an additional company, within two hours.

(3) *Most Probable Course of Action.* The enemy defends CACTF and continues using it as a base of operations and to consolidate weapons and material. His intent is to prevent the capture of CACTF and its support base, provided by its facilities and population.

____
(Classification)

Figure 5-9. Example battalion OPORD for FTX Attack.

(Classification)

b. **Friendly Forces.**

(1) (___) Brigade, (___) Division attacks at (date-time group) to seize the town of CACTF at (grid), to keep the enemy from using it as a base of operations. The intent is to secure the town and the road network in the area, to prevent the enemy from using it as a base of operations.

(2) (___) Battalion, (___) Infantry attacks to seize the road network at OBJ (____), vicinity (grid), at (date-time group) to prevent enemy forces from counterattacking along Route (____).

(3) (___) Battalion, (___) Infantry attacks at (date-time group), to seize the high ground at OBJ (____), vicinity (grid), and to establish defensive positions to block enemy routes in and out of CACTF.

2. **MISSION.** (___) Battalion, (___) Infantry attacks at (date-time group) to seize CACTF, vicinity (grid), NLT (date-time group), to prevent the enemy from using the town as an operations and logistics base.

3. **EXECUTION.**

a. **Concept of the Operation.** The battalion will accomplish its mission, which is to deny the enemy use of CACTF and its surrounding terrain and population, by seizing the town and controlling the key areas in the vicinity. We will conduct a feint in the west and then attack the decisive point within CACTF (OBJ [____]).

(1) *Maneuver.* (___) Company conducts a feint to deceive the enemy, then the main effort (*evaluated*) company seizes OBJ (____), to prevent the enemy from using CACTF as an operational and logistical base. (___) Company supports the main effort by seizing OBJ (____), in order to prevent the enemy from counterattacking the main effort. (___) Company seizes OBJ (____) to support the main effort's attack.

(2) *Fire Support.* Annex C. (See fire support.) Purpose of fires is to suppress the enemy and to obscure movement to CACTF. Priority of artillery fires is to (feinting) Company; on order, priority will shift to (supporting effort) Company. Priority of 81-mm mortar fires is to (*evaluated*) Company.

(3) *Engineering.* The purpose of the engineers is to breach obstacles during the attack, to help (*evaluated*) Company seize OBJ (____). Priority of support is to (*evaluated*) Company, (___) Company, and (___) Company in order. The battalion commander must authorize any booby traps, and point and hasty minefields.

(Classification)

Figure 5-9. Example battalion OPORD for FTX Attack (continued).

(Classification)

b. **Tasks to Maneuver Units.**

(1) *(Evaluated) Company.*

(a) Infiltrate using infiltration lane TANGO.

(b) Seize OBJ (____), vicinity (grid), NLT (date-time group).

(c) Be prepared to perform cordon and search of CACTF, after consolidation and reorganization.

(2) *(Supporting) Company.*

(a) Follow (main attack) Company and infiltrate on infiltration lane TANGO.

(b) Seize OBJ (____), vicinity (grid), NLT (date-time group).

(c) Coordinate the attack with *(evaluated)* Company, so that both units attack at the same time.

(d) Be prepared to perform cordon and search in the objective area after consolidation and reorganization.

(e) Be prepared to assume the mission of *(evaluated)* Company.

(3) *(Supporting) Company.*

(a) Infiltrate using infiltration lane ZEBRA.

(b) Conduct feint to OBJ (____), vicinity (grid), NLT (date-time group).

(c) Be prepared to establish defensive positions vicinity (grid) to prevent enemy movement to and from CACTF.

c. **Tasks to Subordinate Units.**

(1) *(Scout) Platoon.*

(a) During movement, determine location, strength, and disposition of enemy forces in CACTF.

(b) On order, screen the battalion flank from (grid) to (grid) to warn of the approach of enemy units from the southwest

(2) *Mortar Platoon.* Priority of mortar fire is to *(evaluated)* Company. Mortar platoon moves with *(evaluated)* Company, establishes a firing position vicinity (grid), and displaces on order.

(3) *Antitank Platoon.*

(a) Initially, over watch from (grid) to (grid) to secure the unit infiltration.

(b) On order, move vicinity (grid) to support *(evaluated)* Company's consolidation and reorganization.

(Classification)

Figure 5-9. Example battalion OPORD for FTX Attack (continued).

(Classification)

(4) *1/C/58 Engineers.*

(a) Move with (*evaluated*) Company to provide mobility support.

(b) Be prepared to reduce obstacles in CACTF during cordon and search mission.

d. **Coordinating Instructions.**

(1) Report when ready to execute an assault.

(2) Report any movement of enemy forces from the southwest.

(3) No rubbling is authorized, so minimize collateral damage to property.

(4) MOPP0 is in effect.

(5) Terrorist threat warning is red.

(6) Commanders brief back OPORD to battalion commander at battalion tactical CP at (date-time group).

(7) Rules of Engagement (ROE).

(a) Commanders take all steps necessary and appropriate for their units' protection.

(b) The minimum necessary force is used to control the situation.

(c) Leaders must take measures to minimize risk to civilians, without endangering the unit.

(d) Fire is returned directly to its source, not sprayed into a general area.

(e) Firing ceases when the threat is over.

(f) Anyone trying to surrender is allowed to do so.

(g) Civilians and property are treated with respect.

(h) WP can be used vicinity CACTF to aid in isolating the objectives. The requests for indirect fire, within the town, must be authorized by the battalion commander.

4. SERVICE SUPPORT. (See Annex D.)

5. COMMAND AND SIGNAL.

a. **Command.**

(1) The tactical CP moves with (*evaluated*) Company.

(2) The main CP is vicinity (grid) and displaces on order.

(3) The second in command remains in the battalion main CP.

(4) The rear CP is the alternate battalion main CP.

(Classification)

Figure 5-9. Example battalion OPORD for FTX Attack (continued).

(Classification)

b. **Signal.**
(1) Current SOI is in effect.
(2) Radio-listening silence is in effect for all elements except scouts, until prepared to attack.

ACKNOWLEDGE:
JONES
LTC
OFFICIAL:
SMITH
S-3
Annexes: A–Intelligence
B–Operations Overlay
C–Fire Support
D–Service Support

(Classification)

Figure 5-9. Example battalion OPORD for FTX Attack (continued).

(5) The battalion has destroyed the enemy operating in CACTF, and has seized the town. Two companies are defending against a limited counterattack. One company is in a blocking position. The (*evaluated*) company receives a FRAGO (Figure 5-10) to conduct a cordon and search operation. The purpose of the operation is to locate enemy personnel and equipment remaining in the CACTF.

Enemy personnel in squad-sized elements are reported by the civilian population to be hiding in the CACTF. Reports also indicate numerous weapons caches are being hidden by civilians collaborating with the enemy. Battalion has completed its mission and is preparing to conduct a cordon and search.

After completing this, the battalion exfiltrates to perform a passage of friendly lines. (Evaluated) Company moves to perform cordon and search vicinity OBJ (____), beginning (date-time group) to (date-time group), to locate and seize enemy personnel or equipment.

The (*evaluated*) company notifies the commander of the results of the search, and then notifies the commander when they (*evaluated company*) are ready to start movement. The company exfiltrates on order along Route Gold to perform linkup at (grid) and passage of lines at (grid). (See overlay for lanes.)

Figure 5-10. Example FRAGO for FTX Attack.

4. **Support Requirements**.

a. *Minimum Trainers/Evaluators*:. The company commander acts as the trainer-evaluator for the platoons. The battalion may direct the FTX internal evaluation along with a company evaluator and an FO evaluator. The platoon leaders and platoon sergeants are the primary trainers during the initial phases of the training. The company commander, the XO, the platoon leaders, and the platoon sergeants all act as trainers during the company phases of training. If the exercise is conducted by the battalion, either the battalion commander, XO, or S-3 acts as the trainer and evaluator, but additional battalion assets may be required to support the evaluation phase. The controllers or evaluators are positioned at critical locations throughout the exercise to observe the action. They must not interfere with the conduct of the operation, except to halt it when a safety violation occurs or to conduct an AAR.

b. *Vehicles/Communications*: Those organic to the company are used in the exercise. This includes one designated medical evacuation vehicle per platoon and at least one aidman on site.

c. *Opposing Force*:. The OPFOR ground force should be at least platoon-size. FTXs conducted by the company require internal resourcing; units should rotate to ensure total training. OPFOR tasks and standards are provided in each T&EO.

d. *Maneuver Area*: A training area 5 by 10 km which includes the CACTF is best for this exercise.

5. Evaluation.

Table 5-8 lists the T&EOs used to evaluate this STX.

Table 5-8. T&EOs used to evaluate STX Attack.

Task/Function	*Task Number*
Perform Cordon and Search	7-3-1105
Take Action on Contact	7-4-1107
Perform Passage of Lines	7-4-1125
Perform Linkup	7-4-1128
Conduct Tactical Movement	7-2-1342
Occupy Assembly Area	7-4-1136
Maintain Operations Security	7-4-1409
Prepare for Combat	7-4-1606
Conduct Consolidation and Reorganization (Co/Plt)	7-2-5027
Perform Cordon and Search	7-2-1105
Execute an Assault (MOUT)	7-2-1109
Perform Exfiltration	7-1-1146
Perform Infiltration	7-1-1147
Perform Reorganization	7-1-1608
Perform Consolidation	7-1-1609

MECHANIZED INFANTRY COMPANY OR TEAM FTX DEFEND

1. **Objective**. This FTX is designed to train the mechanized Infantry company or team in conducting defensive operations in urban terrain. It also provides platoons an opportunity to prepare for full-scale FTXs conducted and evaluated by the parent battalion. This exercise provides practice for the company commander, platoon leaders, squad leaders, and company headquarters personnel in planning, coordinating, and controlling combat operations.

2. **Interface**. This FTX is supported by the STX C-II-4, C-II-5, and C-II-6 tasks shown in Table 5-9. It is also supported by two warrior drills, *React to Contact* and *Break Contact*.

Table 5-9. Supporting tasks for company FTX.

STX C-II-4	*STX C-II-5*	*STX C-II-6*
Occupy Assembly Area	Perform Ambush	Operate a Checkpoint
Perform Tactical	Establish a Roadblock/Checkpoint	Interdict Smuggling Operations
Road March	Perform CBRN Operations	Defend Convoy
Perform Relief in Place	Conduct Tactical Movement	
Defend Built-Up Area/ Building	Perform Passage of Lines	
	Occupy Assembly Area	

3. **Training**.

a. *Guidance*. FM 7-0 and FM 7-1 provide training guidance, tips, and enhancers for planning and executing STXs. The trainer should review the individual, leader, and collective tasks to be performed during the STX to determine which require initial or refresher training.

b. *General Situation*.

(1) The battalion is ordered to conduct defensive operations in support of the brigade mission; the company is ordered to perform a relief operation and to prepare a defense in an urban area. The company remains in position for at least 36 hours before receiving a follow-on mission. Most civilians have been evacuated from the area, but a few remain. A terrorist threat exists as well as the possibility of enemy battalion-level offensive actions.

(2) This exercise should be conducted under various environmental conditions. The company also conducts operations under threat of CBRN attack.

c. *Special Situation.*

(1) The planning process for the defense mission begins when the company receives a warning order. The exercise ends when the company completes the passage of lines and occupies an assembly area. Figure 5-11 shows a graphic scenario of task performance for the FTX Defend.

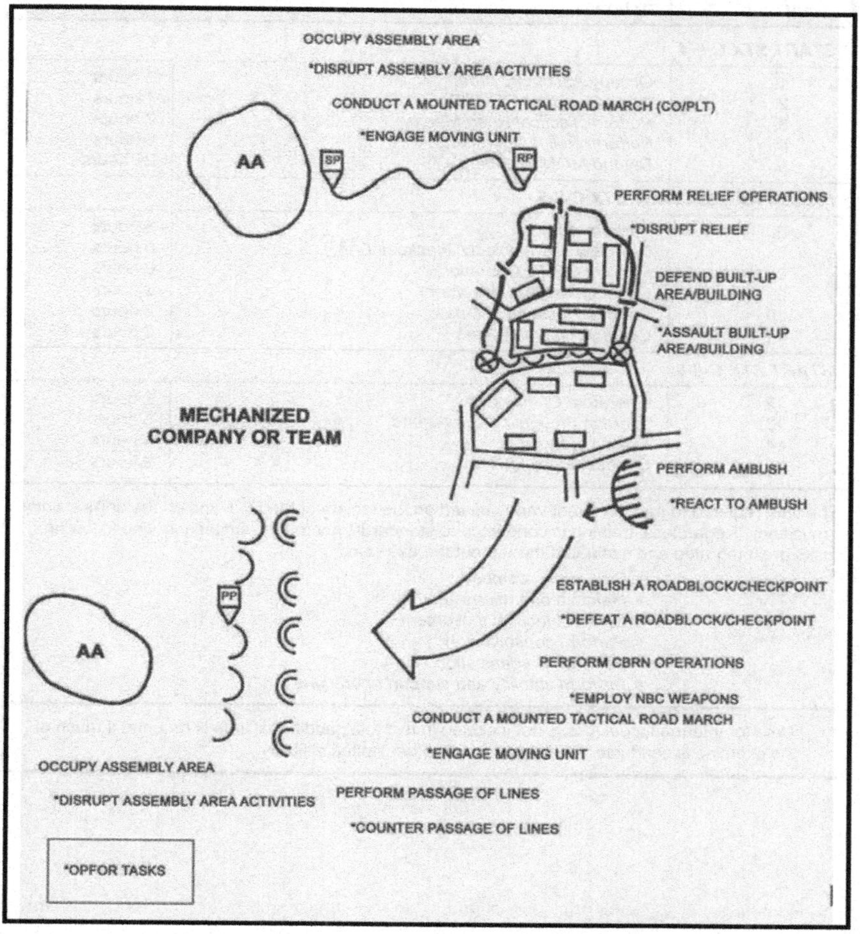

Figure 5-11. Graphic scenario of task performance for FTX Defend.

(2) Table 5-10 provides a recommended sequence for performance of T&EOs from DTMS and the time allotted for each portion of the STX, including AARs.

Table 5-10. Sequence of events and times for FTX Defend.

Event	*Action*	*Estimated Time*
START STX C-II-4		
1.	*Occupy Assembly Area*	2 hours
2.	*Issue Company OPORD*	4 hours
3.	*Perform Tactical Road March*	3 hours
4.	*Perform Relief Operations AAR*	6 hours
5.	*Defend MOUT/Building*	24 hours
ISSUE FRAGO, START STX C-II-5		
6.	*Perform Ambush*	6 hours
7.	*Establish a Roadblock/Checkpoint AAR*	6 hours
8.	*Perform CBRN Operations*	6 hours
9.	*Conduct Tactical Movement*	3 hours
10.	*Perform Passage of Lines*	2 hours
11.	*Occupy Assembly Area*	3 hours
START STX C-II-6		
12.	*Operate a Checkpoint*	2 hours
13.	*Interdict Smuggling Operations*	6 hours
14.	*Defend Convoy*	4 hours
15.	*Conduct Final AAR*	2 hours
The time required to train an event varies based on the factors of METT-T and on the unit's training proficiency. Regardless, training is conducted to standards, not to time limitations. The following tasks are integrated and evaluated throughout the exercise: ▪ *Prepare for combat* ▪ *Maintain operations security* ▪ *Conduct tactical movement* ▪ *Perform consolidation* ▪ *Perform reorganization* ▪ *Perform stability and support operations*		
* Time for intermediate AARs is not included in this total; additional time is required if much of the exercise is conducted at night or during other limited visibility.		

(3) The battalion commander issues the defense OPORD to the company (Figure 5-12).

```
                         ____________
                       (Classification)

                                           Copy no ___ of ___ copies
                                          ___ Battalion, ___ Infantry
                                       Unit location or place of issue
                                      Date-time group (with time zone)
                                            Message reference number
OPERATION ORDER NO. _____

Reference: Map, series no. ___, sheet no. ___, edition ___
Time Zone Used Throughout the Order: ____
TASK ORGANIZATION:
___Co ___Tm ___Tm (-)  Battalion Control
 1/__/__/AR 2/__/__/AR Scout Platoon
    120-mm Mortar Platoon
Team Tank E Co(-) 1/C/1-222 ADA (Stinger) (DS)
1/__/__/Inf  C/58 ENGR (DS)
2/__/__/Inf  1/1 (GSR) TM A/58 MI Bn
    ATK Force B (1/_/_/Inf)
    Battalion Trains

1.  SITUATION.

    a. Enemy Forces. Annex A (see intelligence overlay [prepared by
brigade S-2]).
        (1)  Composition, Disposition, Strength. The company is opposed
by the 1st Dragoons Regiment, 60th Guards Division. This regiment is
at 85 percent strength and has received engineer and artillery
reinforcements. The enemy is equipped with an assortment of former
Soviet and Warsaw pact small arms, machine guns, and armored vehicles
including T-62 tanks.
        (2)  Capabilities. The enemy has established defensive
positions that follow a line from vicinity (grid) to vicinity (grid).
Enemy armored reconnaissance units are operating in the more open
terrain to the southwest of CACTF. The enemy has been observed moving
forces in depth throughout the area in preparation for an offensive
action.
        (3)  Probable Course of Action. The enemy's most probable
course of action will be to begin offensive actions within the next 24
to 36 hours. Enemy encountered includes numerous reconnaissance
elements trying to find the company's vulnerabilities and a main body
trying to exploit any weakness.

                         ____________
                       (Classification)
```

Figure 5-12. Example battalion OPORD for mechanized company or team STX Defend.

(Classification)

b. **Friendly Forces.**

(1) (___) Brigade, (___) Division relieves (___) Brigade in place at (date-time group), to defend in sector from (grid) to (grid) to (grid) to (grid), in order to secure the corps counterattack axis of advance, (grid). The intent is to secure the road network around CACTF to prevent the enemy from impeding the movement of the (___) Brigade, (___) Division.

(2) TF (__) is the main effort in the defense and, NLT (date-time group), holds key terrain vicinity (grid) on our left, to prevent an enemy bypass of the brigade left flank.

(3) TF (__), on our right, defends along Route 360, vicinity (grid) to (grid) to (grid) to (grid), NLT (date-time group), to prevent an envelopment of the right flank. They conduct a supporting attack at (date-time group) to seize the high ground at OBJ OAK, vicinity (grid), to establish defensive positions that prevent the enemy armored reconnaissance units from interfering with the seizure of the pass at (grid) or the attack of the (___) Battalion, (___) Infantry, Mechanized.

2. **MISSION.** TF (____) defends NLT (date-time group), to retain the town of CACTF and its surrounding road network and allow brigade to prevent enemy interference with corps counterattack.

3. **EXECUTION.**

a. **Concept of the Operation.** Annex B (see operations overlay).

(1) Maneuver. Company (main effort) retains the main intersection (traffic loop) in CACTF, to prevent enemy movement through the sector. Team (*evaluated*) blocks movement west of 3d Street, to prevent the enemy from enveloping the main effort. The counterattack force is committed as needed. Team (__) blocks movement east of Oak Street, to canalize the enemy into EA (____). Team (__) defends from BP1, to prevent a coordinated attack against the main effort. E Company occupies BPs, to destroy enemy vehicles and protect the TF flanks.

(2) Fire Support. (See Annex C.) Fires are used to suppress the enemy and obscure our movement during the relief operation. Once positions are occupied, fires are used to suppress enemy overwatch positions, disrupt attempts to breach obstacles, and separate enemy Infantry from their supporting carriers. Priority of CAS and indirect fire is initially to Team (*evaluated*), during the relief operation, and then to Team (main effort). Teams (main effort) and (*evaluated*) each have two priority 155-mm targets. Each team may make one priority target an FPF.

(Classification)

Figure 5-12. Example battalion OPORD for mechanized company or team STX Defend (continued).

____
(Classification)

(3) Counter-Air Operations. Priority of protection is to the TF trains and main CP. Air defense weapons status is free; air defense warning status is yellow.
(4) Intelligence. Priority of collection is to identify and locate the C2 elements of the MRR and MRB, enemy tank battalions, and the enemy assault detachments' actual axes of advance.
(5) Engineering. The purpose of obstacles is to turn the enemy into restrictive terrain and engagement areas. Priority of support is to Team (main effort), Team (*evaluated*), Team (__), and Team (__), in that order.

b. **Tasks to Maneuver Units**.
(1) Team (Main Effort).
(a) Sector of defense includes the east side of Front Street, the south side of 1st Street, the west side of 3d Avenue area, and the north side of 3d Street.
(b) Team (main effort) relieves Team (__), TF (__).
(2) Team (Evaluated).
(a) Sector includes the east side of 3d Avenue, the south side of 1st Street, the west side of 6th Avenue, and the north side of 3d Street.
(b) Team (*evaluated*) relieves Team (__), TF (__).
(3) Team (__).
(a) Sector includes the east side of Oak Street, the south side of 1st Street, the east side of Front Street, and the north side of 3d Street.
(b) Team (__) relieves Team (__), TF (__).
(c) Team (__) coordinates directly with right flank team of TF (__), to ensure mutually supporting fires.
(4) Team (__)
(a) Occupy BP 1 vicinity (grid)
(b) Be prepared to occupy BP 2 (grid) to engage enemy in EA (_____).
(5) Company E.
(a) Occupy BP 2-1 (grid) and move on order to BP 2-2 (grid).
(b) Provide antiarmor fires to destroy enemy tanks in EA (_____).
(6) Counterattack B (Reserve).
(a) Occupy AA Black, vicinity (grid).
(b) Reconnoiter routes into TF (main effort) sector.

____
(Classification)

Figure 5-12. Example battalion OPORD for mechanized company or team STX Defend (continued).

(Classification)

(3) Counter-Air Operations. Priority of protection is to the TF trains and main CP. Air defense weapons status is free; air defense warning status is yellow.

(4) Intelligence. Priority of collection is to identify and locate the C2 elements of the MRR and MRB, enemy tank battalions, and the enemy assault detachments' actual axes of advance.

(5) Engineering. The purpose of obstacles is to turn the enemy into restrictive terrain and engagement areas. Priority of support is to Team (main effort), Team (*evaluated*), Team (__), and Team (__), in that order.

b. **Tasks to Maneuver Units.**

(1) Team (Main Effort).

(a) Sector of defense includes the east side of Front Street, the south side of 1st Street, the west side of 3d Avenue area, and the north side of 3d Street.

(b) Team (main effort) relieves Team (__), TF (__).

(2) Team (Evaluated).

(a) Sector includes the east side of 3d Avenue, the south side of 1st Street, the west side of 6th Avenue, and the north side of 3d Street.

(b) Team (*evaluated*) relieves Team (__), TF (__).

(3) Team (__).

(a) Sector includes the east side of Oak Street, the south side of 1st Street, the east side of Front Street, and the north side of 3d Street.

(b) Team (__) relieves Team (__), TF (__).

(c) Team (__) coordinates directly with right flank team of TF (__), to ensure mutually supporting fires.

(4) Team (__)

(a) Occupy BP 1 vicinity (grid)

(b) Be prepared to occupy BP 2 (grid) to engage enemy in EA (____).

(5) Company E.

(a) Occupy BP 2-1 (grid) and move on order to BP 2-2 (grid).

(b) Provide antiarmor fires to destroy enemy tanks in EA (____).

(6) Counterattack B (Reserve).

(a) Occupy AA Black, vicinity (grid).

(b) Reconnoiter routes into TF (main effort) sector.

(Classification)

Figure 5-12. Example battalion OPORD for mechanized company or team STX Defend (continued).

(Classification)

(c) Priority for planning:
- Battle position to reinforce main effort.
- Battle position to counterattack into Team (main effort) sector.
- Battle position to counterattack into Team (*evaluated*) sector.

(1) Scout Platoon.

(a) Determine location, strength, and disposition of enemy forces.

(b) Perform a screen of the TF's right flank from (grid) to (grid) beginning (date time group) to prevent enemy infiltration of CACTF.

c. **Tasks to Subordinate Units.**

(1) Mortar Platoon.

(a) Priority of fire to TF (main effort). Move with TF (main effort) and establish firing position vicinity (grid).

(b) Displace on order to (grid).

(2) 1/C/1 222 ADA. Coordinate locations with S-3 NLT (date time group).

(3) C/58 Engineers.

(a) Coordinate the engineer effort with the team commanders NLT (date time group).

(b) Provide an obstacle clearing team to Counterattack Force B, NLT (date time group).

(4) 1/1 (GSR) TM A/58 MI. Coordinate location of assets with the task force S-3 NLT (date time group).

d. **Coordinating Instructions.**

(1) Selective rubbling is authorized, but destruction of more than one building requires TF commander approval.

(2) Destruction of subterranean approaches requires TF commander approval.

(3) Commanders brief back OPORD to TF commander at the TF tactical CP, vicinity (grid), at (date time group).

(4) MOPP1 is in effect as of (date time group).

(5) Decontamination site established vicinity Miller Pond (grid).

(Classification)

Figure 5-12. Example battalion OPORD for mechanized company or team STX Defend (continued).

(Classification)

4. SERVICE SUPPORT. (See Annex D.)

a. **General.**
(1) Combat trains located vicinity (grid).
(2) Field trains located in the BSA, vicinity (grid).

b. **Materiel and Services.**
(1) Supply.
(a) Class I. Ration cycle C-A-C. Team establishes MRE cache sufficient for 48 hours in sector.
(b) Class III. Resupply available on request. Class III located in combat trains.
(c) Class IV. Materials are allocated as shown in Annex D.
(d) Class V. Each team establishes ammunition caches in sector, but should also have two basic loads of ammunition on hand.
(2) Transportation. The main streets are capable of two way heavy vehicle traffic (class 60). Secondary streets may be one way and of lighter construction (class 50).
(3) Maintenance.
(a) Priority by unit: Team (main effort), Team (*evaluated*), Team (__), Company E, Team (__).
(b) Priority by vehicle: M1A1, M2/M3, M113.

c. **Medical Evacuation and Hospitalization.**
(1) TF aid station located vicinity (grid).
(2) Civilian casualties should be transported by nonmilitary means to clinic, located vicinity (grid).

d. Personnel.
(1) No replacements for the next 96 hours.
(2) EPW collection point located vicinity (grid).
(3) Mortuary affairs point located in BSA, vicinity (grid).

e. **Civil-Military Cooperation.**
(1) Curfew in effect 1800 to 0800 daily.
(2) Majority of civilians have been evacuated. Those remaining have been notified of the risk involved.

(Classification)

Figure 5-12. Example battalion OPORD for mechanized company or team STX Defend (continued).

(Classification)

5. **COMMAND AND SIGNAL.**

a. **Command.**
(1) Brigade main CP located vicinity (grid).
(2) TF tactical CP located vicinity (grid).
(3) TF main CP located vicinity (grid).
(4) TF commander located at tactical CP.
(5) TF XO located at main CP.

b. **Signal.**
(1) SOI index 1 9 in effect.
(2) Primary means of communication is wire, FM, messenger in that order.
(3) Building and obstacle marking IAW TF SOP.

ACKNOWLEDGE:

WILLIAMS
LTC

OFFICIAL:

SMITH
S-3

Annexes: A–Intelligence
B–Operations Overlay
C–Fire Support
D–Service Support

(Classification)

Figure 5-12. Example battalion OPORD for mechanized company or team STX Defend (continued).

(3) The TF has retained the north half of the CACTF after the enemy attack. The enemy was defeated but has established squad-sized to platoon-sized strongpoints in the southern half of the CACTF. The enemy is expected to receive reinforcements within the next 12 to 24 hours. Theprevent enemy reinforcements from moving to the strongpoints to strengthen and resupply them, TF headquarters orders the team to conduct platoon-sized ambushes. They also order the team to prepare to establish a roadblock, to control refugee flow through the area of operations, before the team moves out of the area. Headquarters issues a FRAGO (Figure 5-13).

```
Numerous enemy strongpoints are located to your front, vicinity (grid), (grid), and (grid). We expect these strongpoints to receive reinforcements within the next 12 to 24 hours.

Your mission is to establish platoon ambushes on the likely avenues of approach to the strongpoints, NLT (date-time group), in order to prevent reinforcements or resupply to these strongpoints.

Notify the TF commander when you are prepared to move.

Coordinate with TF scouts for guides. On order, after return from ambush missions, establish roadblocks at (grid) and (grid). Be prepared to move within the next 36 to 48 hours to conduct further follow-on missions.
```

Figure 5-13. Example FRAGO for FTX Defend.

4. **Support Requirements.**

a. *Minimum Trainers/Evaluators*: This exercise should be conducted for practice with the company commander as the trainer-evaluator for his platoons. The battalion may direct the FTX (internal evaluation) using a company evaluator and an FO evaluator. The platoon leaders and platoon sergeants are the primary trainers during the initial phases of the training. The company commander, the XO, the platoon leaders, and the platoon sergeants act as trainers during the company phases of training. If the exercise is conducted by the battalion, either the battalion commander, the XO, or the S-3 acts as the trainer and evaluator, but may require additional assets from the battalion to support the evaluation phase. The controllers or evaluators must be positioned at critical locations throughout the exercise, to observe the actions. They must not interfere with the conduct of the operation, except to halt it when a safety violation occurs or to conduct an AAR.

b. *Vehicles/Communications*: Those organic to the company are used.

c. *Opposing Force*:. The OPFOR ground force should be at least platoon-size. FTXs conducted by the company require internal resourcing, and units should be rotated to ensure total training. T&EOs provide training tasks and standards for the OPFOR.

d. *Maneuver Area*: A training area at least 5 by 15 km is best for this exercise.

e. *Consolidated Support Requirements*: This exercise requires the items shown in Table 5-11.

f. *Barrier materials.* Sufficient Class IV must be made available to fortify positions, IAW FM 3-06.11.

Table 5-11. Ammunition and pyrotechnics support requirements for a six-day FTX.

Ammunition	*DODAC*	*Per Plt/Co/Bn FTX*
5.56-mm blank	1305A080	120 rounds per rifle
5.56-mm SESAMS		120 rounds per rifle
5.56-mm blank	1305A075	200 rounds per M249 LMG
5.56-mm SESAMS		200 rounds per M249 LMG
7.62mm	1305A111	200 rounds per BFV/M1
		400 rounds per M60/240B
ATWESS cartridge	1370L367	6 (Viper) (for LAW) 18 per Dragon 14 per TOW system 9 per OH-58D/AH-64
Pyrotechnic device	1370L602	6 per M1 2 per target
Hand grenade fuze (M228)	1330G878	150 per battalion
Stun grenade M84	1330GG09	150 per battalion
Smoke grenade, HC		150 per battalion
Smoke pot		6 per battalion

5. **Evaluation**.

Table 5-12 lists the DTMS T&EOs used to evaluate this STX.

Table 5-12. T&EOs used to evaluate STX Attack.

Task/Function	*Task Number*
Defend MOUT/Building	7-3/4-1118
Conduct a Dismounted Tactical Road March (Co/Plt)	7-2-1189
Conduct a Mounted Tactical Road March (Co/Plt)	7-2-1198
Perform Relief Operations	7-3/4-1124
Perform Passage of Lines	7-3/4-1125
Conduct Tactical Movement	7-2-1342
Occupy Assembly Area	7-3/4-1136
Establish a Roadblock/Checkpoint	7-3/4-1401
Perform CBRN Operations	7-3/4-1406
Maintain Operations Security	7-3/4-1409
Prepare for Combat	7-3/4-1606
Conduct Consolidation and Reorganization	7-2-5027
Perform Ambush	7-2-1106

TARGETRY

5-10. Table 5-13 lists the combination of precision and nonprecision targets to support the diversity of scenarios required to support combined arms and branch specific lane training. Precision targets, less than 50 meters, are used in and around buildings within the CACTF and reinforce precision marksmanship techniques described in FM 3-22.9 (Table 5-13).

Table 5-13. CACTF target requirements.

Event	*Number Of Lanes*	*Human Targets*		*Vehicle Targets*	*Totals*	
					HT	*VT*
Company/Platoon STX Offense	3/9	Precision	Nonprecision	9	45	9
		15	30			

CONSIDERATIONS FOR TRAINING WITH ARMOR

5-11. Built-up areas consist mainly of man-made features such as buildings, streets, and subterranean systems. These features of urban terrain create a variety of tactical problems and possibilities. In order to ensure that the tank platoon can operate effectively in the urban environment, the platoon observation and direct-fire plans must address the ground-level or surface fight (in streets and on the ground floor of buildings), the aboveground fight (intrasurface and supersurface), and the subterranean fight (Figure 5-14). Infantry can assist the tank platoon by—

- Locating targets for tanks to engage.
- Destroying antitank weapons.
- Assaulting enemy positions and clearing buildings with tank support.
- Protecting tanks from antitank fires.

Figure 5-14. Urban battlespace.

TRANSPORTING INFANTRY

5-12. At times, the tank platoon must transport Infantrymen (Figure 5-15). They only do so when not expecting contact. If the platoon is moving as part of a larger force, and if it is tasked to provide security for the move, then the lead section or element does not carry Infantry.

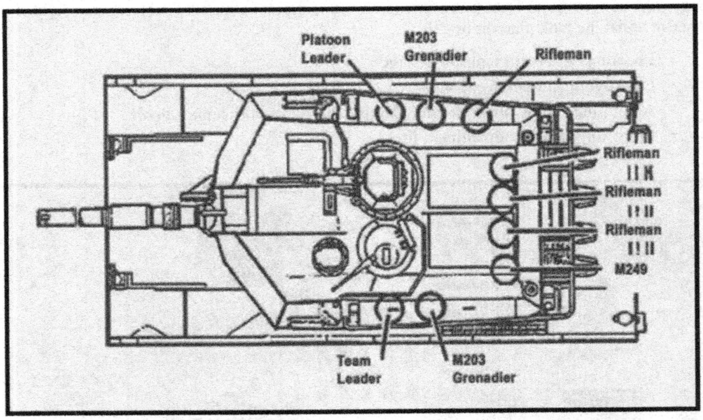

Figure 5-15. Example positions for Infantry riding on a tank.

5-13. Infantry and Armor leaders must observe the procedures, precautions, and considerations in Figure 5-16 when Infantrymen ride on tanks:

- Infantry should thoroughly practice mounting and dismounting procedures and actions on contact.
- Infantry must always alert the TC before mounting or dismounting. They must follow the commands of the TC.
- Infantry platoons should be broken down into squad-sized groups, similar to air assault chalks, with the Infantry platoon leader on the Armor platoon leader's vehicle and the Infantry PSG on the Armor PSG's vehicle.
- Platoon leaders, PSGs, and team leaders should position themselves near the TC's hatch, using the external phone (if available) to talk to the TC and relay signals to the unit.
- If possible, the lead vehicle should not carry Infantrymen. Riders restrict turret movement and are more likely to be injured or killed on initial contact.
- Whenever possible, Infantrymen should mount and dismount over the left front slope of the vehicle. This procedure ensures that the driver can see the Infantrymen and that the Infantrymen DO NOT pass in front of the coax machine gun. Infantrymen must ensure that they remain behind the vehicle's smoke grenade launchers. This will automatically keep them clear of all weapon systems.
- Infantrymen must always have three points of contact with the vehicle; they must watch for low-hanging objects such as tree branches.
- All Infantrymen should wear hearing protection.
- Infantrymen should not ride with anything more than their battle gear. Rucksacks and B-bags should be transported by other means.
- Infantrymen should scan in all directions. They may be able to spot a target the vehicle crew does not see.

Figure 5-16. Precautions for Infantrymen riding on a tank.

5-14. Passengers should be prepared to take the actions on contact shown in Figures 5-17 and Figure 5-18.

- Wait for the vehicle to stop.
- At the TC's command, dismount immediately (one fire team on each side). ***Do not*** move forward of the turret.
- Move at least 5 meters to the sides of the vehicle.
  Do not move behind or forward of the vehicle.
- ***Do not*** move in front of vehicles unless ordered to do so. Main gun discharge overpressure can inflict severe injury or death to forward dismounts.
- ***Do not*** dismount a vehicle unless ordered or given permission to do so.
- ***Do not*** dangle arms or legs, equipment, or anything else off the side of a vehicle; they could get caught in the tracks, causing death, injury, or damage to the equipment or vehicle.
- ***Do not*** place too many riders on the vehicle.
- ***Do not*** fall asleep when riding. The warm engine may induce drowsiness; a fall could be fatal.
- ***Do not*** smoke when mounted on a vehicle.
- ***Do not*** stand near a moving or turning vehicle at any time. Tanks have a deceptively short turning radius.

Figure 5-17. Actions on contact for Infantrymen riding on a tank.

DANGER

1. The overpressure from the tank 120-mm cannon can kill a dismounted Infantryman within a 90-degree arc extending from the muzzle of the gun tube out to 200 meters.
2. From 200 to 1,000 meters along the line of fire, on a frontage of about 400 meters dismounted Infantry must be aware of the danger from discarding sabot petals, which can kill or seriously injure Soldiers.

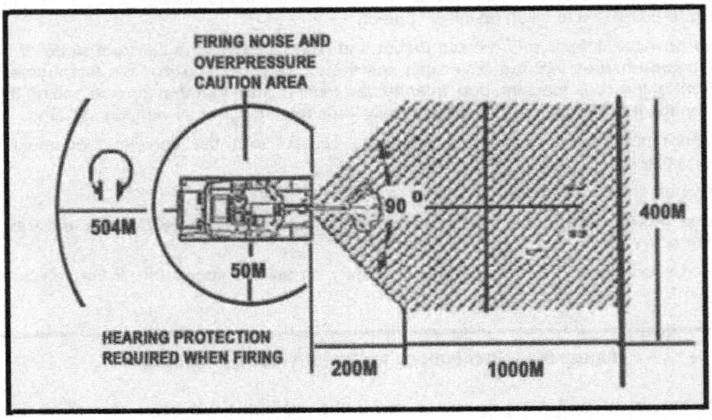

Figure 5-18. Danger areas around a tank when firing the 120-mm main gun.

ADDITIONAL CONSIDERATIONS

5-15. Additional considerations and preparations for transporting Infantrymen include the following:

- Armor--
  - Uses main gun fire to reduce obstacles or entrenched positions for the Infantry.
  - Takes directions from the Infantry ground commander (platoon leader/PSG/squad leader) to support their fire and maneuver.
  - Provides reconnaissance by fire for the Infantry.
  - -- Should know and understand how the Infantry clears buildings, how they mark cleared buildings, casualty evacuation plan, signal methods, engagement criteria for tank main gun, front line trace reporting, ground communication from the tank with the dismounted Infantry.
  - -- Uses its night vision capability to augment and supplement the Infantry's night vision capability.
- Infantry--.
  - -- Provides real time information for the tank crewmen to help them overcome tank noise and the lack of ground situational understanding.
  - -- Provides reconnaissance and fire direction of enemy positions for main gun attack.

- Considerations for dismounted tank security include--
  -- Tank crewman should rehearse the mounting and dismounting of Infantry from their vehicle, briefing the Infantrymen on safety procedures for the vehicle and weapon systems.
  -- Tank commanders need to rehearse communicating with dismounted Soldiers by way of TA-1 and DR-8 in the bustle rack.
- Vehicle preparation for combat in urban terrain should cover--
  -- Keep at least one ballistic shield to the "Dog House" closed. Most engagements will be under boresight range, and the battlesight technique will suffice.
  -- Place sandbags around antenna connections and electrical wiring on the turret top.
  -- Place extra coax ammunition inside the turret.
  -- Remove any highly flammable products from the outside of the vehicle and from the sponson boxes.

VEHICLES, WEAPONS, AND MUNITIONS

5-16. Numerous factors related to vehicles and their organic weapons and munitions affect the tank platoon's urban operation planning and execution, including the following:

- The preferred main gun rounds in the urban environment are HEAT, MPAT (ground mode), and MPAT-OR. These all perform much better than sabot rounds against bunkers and buildings.
- HEAT ammunition will open a larger hole in reinforced concrete or masonry structures than MPAT or MPAT-OR. Both MPAT and MPAT-OR, however, offer greater incapacitation capability inside the structure.
- HEAT ammunition arms approximately 60 feet from the gun muzzle. It loses most of its effectiveness against urban targets at ranges of less than 60 feet.
- MPAT and MPAT-OR rounds arm approximately 100 feet from the muzzle of the gun. Because of the shape and metal components of the projectiles, however, this ammunition remains effective at ranges of less than 100 feet.
- Sabot petals, including those on MPAT and MPAT-OR, endanger accompanying Infantry elements. They create a hazard area extending 70 meters on either side of the gun-target line out to a range of 1 kilometer.
- The tank's main gun can depress only to -10 degrees and can elevate only to +20 degrees. This creates considerable dead space for the crew at the close ranges that are typical in the urban environment.
- The external M2 HB machine gun can deliver a heavy volume of suppressive fire and penetrate light construction, buildings and most barricades. The M2 HB MG can elevate to +36 degrees; however, the TC must be unbuttoned to fire the M2 on the M1A2 or M1A2 SEP.
- The M240 coax machine gun can effectively deliver suppressive fires against enemy personnel and against enemy positions that are behind light cover.
- The loader's M240 machine gun can effectively deliver suppressive fire against enemy personnel and against enemy positions that are behind light cover; however, the loader must be unbuttoned to operate it. This weapon may be dismounted and used in a ground role if units are equipped with the M240 dismount kit.

- When buttoned up, the tank crew has limited visibility to the sides and rear and no visibility to the top. Figures 5-19 and 5-20 show the dead spaces associated with tank operations in an urban environment.
- FM 3-20.12 explains special uses for tank-mounted machine guns in the urban environment.

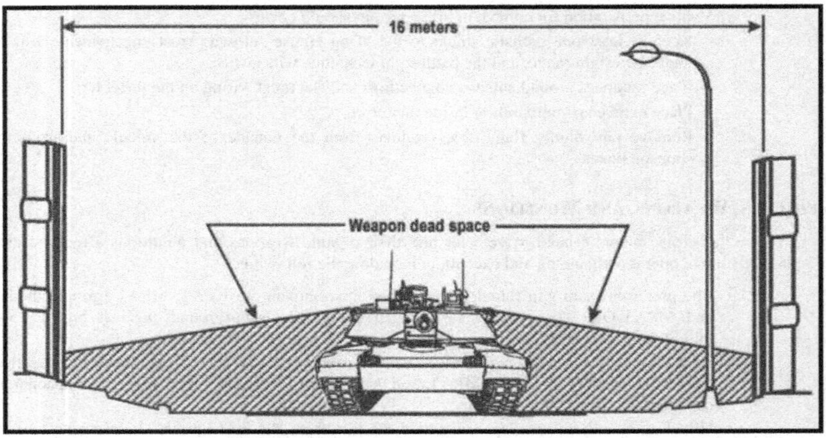

Figure 5-19. Tank weapon dead space at street level.

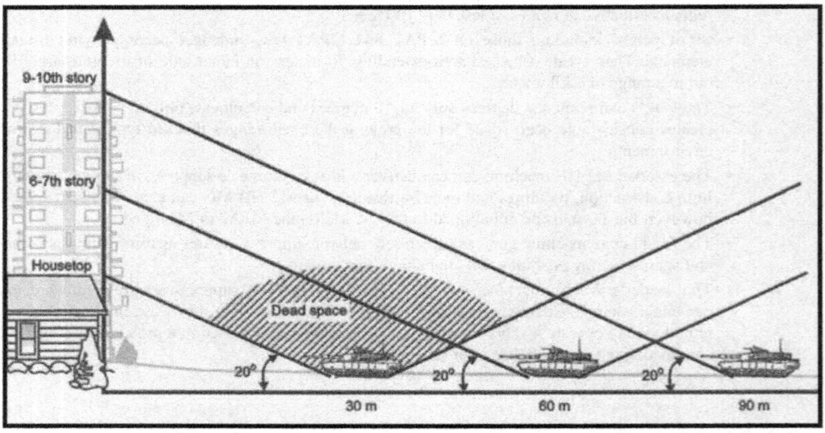

Figure 5-20. Tank main gun and coax dead space above street level.

TANK PLATOON COMMAND AND CONTROL IN URBAN OPERATIONS

5-17. Communications problems, fire control, proximity and visibility, personnel factors, ROE, ROI, civilians, and slow-pace urban operations all affect the tank platoon's urban operations planning and execution:

Communications Problems

5-18. The low-level task organization that may take place during urban operations requires elements to establish additional communications links, which can be disrupted by buildings and other urban terrain features.

Fire Control

5-19. Extensive direct fire planning and restrictive fire control measures are an absolute requirement in urban operations.

Proximity and Visibility

5-20. Friendly elements often must operate in confined and restrictive areas during urban operations, and they may not be able to see other nearby friendly forces. These factors significantly increase the danger of fratricide.

Personnel Factors

5-21. Urban operations impose significant, and often extreme, physical and psychological demands on Soldiers and leaders.

ROE, ROI, and Civilian Considerations

5-22. The ROE or ROI may restrict the use of certain weapon systems and TTP. As an integral part of urban operations, noncombatants create special operational problems. Thedeal with these concerns, units operating in urban terrain must know how to effectively employ linguists and counterintelligence and civil affairs teams.

Slow Pace Urban Operations

5-23. This operation usually prevents the tank platoon from taking full advantage of the speed and mobility of its tanks.

Appendix A

Training Tips

The effectiveness of urban operations training depends on the imagination and thoroughness applied to a training plan. All trainers, from training managers to junior leaders, should prepare their units to fulfill their urban operations mission. They can employ procedures that are based on training doctrine and proven techniques. This appendix discusses several techniques for conducting urban operations training. Trainers make the choice based on the training needs and available resources.

CONCURRENT TRAINING

A-1. Concurrent training allows groups of Soldiers to train on different tasks at the same time. This training requires a plan for rotating Soldiers between the stations that teach different tasks. This procedure is referred to as round-robin training. Concurrent training is an efficient training technique since it makes the best use of training time and facilities. Concurrent training in the urban operations training facilities works best with performance-oriented training. It also requires detailed planning, control, coordination, and assistant trainers. Each station should conduct training that is relevant to, and supports the training the unit is conducting.

URBAN ASSAULT COURSE AND BREACH FACILITY

A-2. These facilities are well suited for conducting concurrent training. After mastering the tasks at one station, a group of Soldiers rotates to the next station in the sequence. Theavoid bottlenecks at any of the stations, trainers plan instructional blocks that last about the same length of time. Concurrent training is often conducted with dedicated instructors at each station. The *advantage* of this technique is that it ensures uniformity of performance standards and saves on trainer preparation time. The *disadvantage* is that the Soldier is not always trained by his leader. Therefore, when feasible, small-unit leaders should rotate with their Soldiers from station-to-station so they can take part in the training. This technique demands prior planning, preparation, and instructors that can teach.

URBAN ASSAULT COURSE AND COMBINED ARMS COLLECTIVE TRAINING FACILITY

A-3. These facilities are well suited for conducting concurrent exercises. Concurrent exercises are variants of concurrent training. In a concurrent exercise, elements of the same unit receive different missions. These missions correspond to each element's training objective for that exercise. This technique can be used during FTXs, STXs, or CFXs. For example, one unit attacks; a second unit supports by fire; a third unit acts as a reserve. Also, elements of the same unit can challenge each other. For example, if a trainer has been training his unit in urban offense and defense operations, he can create a scenario in which one force is required to attack the other. This technique allows the unit to train with an OPFOR without an extra investment of resources. Leaders must remember to carefully plan the force ratios between the OPFOR and counterattacking forces.

DEMONSTRATIONS

A-4. Trainers must demonstrate the correct way to perform the tasks that the Soldiers will be required to perform. A demonstration is more effective when the trainer uses hand-on techniques. A live demonstration is an effective teaching method since it holds Soldiers' interest and they are able to learn with their eyes. The UAC and the Breach Facility are ideal facilities for teaching individual and collective tasks using demonstrations. The trainer must ensure that each Soldier. Performs the task being taught to guarantee that the soldier has learned. Key points should be explained to the satisfaction of each Soldier before progressing to the next step. The trainers can use video cameras to film the action occurring inside the buildings to show the Soldiers the correct way to perform a task or use the same action to explain a better technique to accomplish the same task. This technique is very effective because the Soldiers can watch the demonstrations on a remote monitor, and trainers can replay the videotape as often as required to reinforce his training objective.

EXERCISES

A-5. Training exercises are used to develop tactical SOPs, teamwork among the participants, and Soldier confidence in his leaders and equipment. Table A-1 matches training exercise types to appropriate training facilities.

Table A-1. Training exercise matrix.

EXERCISE	*CACTF*	*SHOOT HOUSE*	*SIMULATIONS*	*LOCAL TOWN*
MAPEX	X		X	X
TEWT	X			X
CPX	X		X	X
CFX	X			X
BSX			X	
LFX		X		
STX	X	X		X
FTX	X			X

Map Exercise

A-6. A MAPEX allows participants to plan and execute urban operations using maps, overlays, sand tables, photographs, or models. MAPEXs develop teamwork and skills in terrain analysis, planning, coordination, order writing, and employment of urban operation tactics and are more suited for battle staff and leaders.

- Using a model of the CACTF is an excellent training aid for instructing Soldiers and leaders. Gather leaders around the map or model and have them devise a tactical plan. Use figures or markers to represent friendly units and the OPFOR. Create tactical situations that challenge participants to react as they would in real combat.
- Incorporate direct fire, casualties, CBRN, logistics, and civilians into the exercise. Ensure the ROE for each exercise is clear to participants.
- Have plenty of detailed maps available for use in the MAPEX and in the planning and execution of subsequent CFX operations.

Tactical Exercise Without Troops

A-7. A TEWT focuses on leader and staff tasks. The senior trainer selects the terrain for the urban TEWT. Either the CACTF or a real village, town, or city is appropriate. The procedure for conducting a TEWT is as follows:

- Create a scenario and enemy situation, and assign missions.
- Allow the leaders to devise their reconnaissance plans and implement them on the terrain.
- Require the leader to brief subordinates while walking the terrain.
- Ensure their plans include the following tasks:
  -- Analyze the terrain.
  -- Employ units according to the terrain analysis.
  -- Emplace weapons.
  -- Devise a unit plan.
- Conduct an AAR.

Command Post Exercise and Battle Simulation Exercise

A-8. CPXs and BSXs allow leaders and battle staffs to perform their tasks in a unit CP, where they receive information as reports requiring certain reactions. This type of training can be achieved through the constructive simulations training portion of the urban operations training strategy using JANUS or JCATS. A company CP is usually associated with a battalion level or higher CPX, but a company can conduct its own CPX. The headquarters section can set up a CP in a CACTF building, displace to another building, and repeat the exercise until they meet the required standards. The section can practice hiding wire on pavement, stringing it between buildings, and running it through sewers. The CP can practice receiving reports, posting situation maps, and sending reports forward to higher headquarters. The CPX/BSXs are most effective in achieving the following general training objectives:

-- Preparing plans.
-- Issuing orders.
-- Building teams.
-- Reconnoitering, selecting, and tactically occupying CP locations.
-- Establishing and using communications.
-- Preparing and sending reports.
-- Coordinating between subordinate elements.

Command Field Exercise

A-9. The CFX is used to train commanders, staffs, and junior leaders. The CACTF or a local town can accommodate this type of exercise. However, the unit commander must go through the correct procedures when conducting training off federal property. The CFX employs all of a unit's command and control assets, with a number of subordinate elements reduced according to a set scale. For example, one Soldier might represent an entire squad, one vehicle a platoon, or one gun an entire battery.

Preliminary Training

A-10. Preliminary training, such as CPXs, TEWTs, MAPEXs, and BSXs, should precede the CFX. Units conduct operations, and controllers resolve battle outcomes. The exercise ends with an AAR. Fewer controllers are needed in a CFX than in an FTX. However, if the number of unit functions increases, the number of controllers must increase also.

Planning Procedures

A-11. Combat in urban areas entails centralized planning with decentralized execution.

Centralized

A-12. Centralized planning can be practiced in a CFX; however, it can be practiced more efficiently in a CPX, TEWT, MAPEX, or BSX.

Decentralized

A-13. Decentralized execution means that platoon, squad, and team leaders lead the battle in an urban area. However, because element participation is scaled down in a CFX, trainers usually find STXs, FTXs, and TEWTs more useful.

LIVE-FIRE EXERCISE

A-14. Chapter 3 and local range regulations provide specific guidance for conducting LFXs at the shoot house. Live-fire training allows Soldiers and leaders to maneuver while firing service ammunition or SRTA. An LFX is most productive when preceded by non-live-fire practice (dry fire or dry run). However, dry fire should *always* precede live fire, with either TES or SESAM. An AAR should follow each repetition, and units should only proceed to the next level of difficulty after mastering the current step. The training in an LFX occurs in the following sequence, which can be shortened if time and ammunition dictate.

1. Dry fire to practice unit drills and individual tasks.
2. Conduct an AAR.
3. Practice drills, tasks a second time with limited use of ammunition to show the complex coordination of fire and maneuver required.
4. Conduct an AAR.
5. Practice exercise a third time with full use of ammunition to enhance realism and build confidence.
6. Conduct an AAR.

SITUATIONAL TRAINING EXERCISE

A-15. STXs are mission-oriented, limited exercises used to train a unit to perform a single collective task or a series of related tasks and drills. With their distinct start and stop points, STXs represent a segment of battle. An STX is more flexible and complex than a single T&EO or drill. In fact, an STX typically includes drills, leader tasks, and separate individual tasks, and involves a unit's "slice" of external combat assets. An STX often precedes an FTX or CFX, and can be conducted in the CACTF by creating a brief situation and mission. For example, the MP company commanders might want to teach his Soldiers about straggler control. They can string the tasks together into small scenarios such as STXs, describe a tactical situation that produces stragglers from nearby units, use other Soldiers to represent the stragglers, and train Soldiers to process stragglers at straggler collection and transport points.

FIELD TRAINING EXERCISE

A-16. FTXs allow units to execute collective tasks under conditions that closely resemble actual urban terrain. FTXs are suitable for the CACTF.

- An FTX requires significant time, resources, and planning. The time is well spent if units master their collective tasks. An FTX is an excellent way to conduct the *run* stage of collective training. Additionally, OPFOR and MILES greatly enhance the training value of the FTX.
- A unit's urban operations FTX training plan addresses ROE, maneuver restrictions, and the controller plan. Soldiers must know the exercise's training objectives before they begin, so they can focus on the correct tasks. Every participant, regardless of position or rank, must fully understand all safety requirements.

- To make the most of their training time in the field, leaders must plan in great detail. The concept of multiechelon training demands that they consider what training can occur at each level in each event.

EVALUATORS

A-17. Soldiers from the training unit or from another unit can evaluate urban operations training, especially during the *run* stage of training in an STX, FTX, or CFX. A dedicated evaluator provides objective and thorough comments on the unit's performance throughout all stages of the exercise. Evaluators monitor the critical point of the action but must avoid interfering. They must not disrupt unit performance or undermine realism. They must be trained and given standardized evaluation guides if their feedback is to be meaningful; MTPs and T&EOs are a good start. Evaluators facilitate AARs alone or along with the unit leader.

TACTICAL ENGAGEMENT SYSTEM

A-18. The TES realistically replicates the effects of actual fire in training. It uses laser adapters on the weapons and laser-sensitive equipment on the Soldiers and vehicles. A TES weapon emits a laser beam when fired. Accurate fire strikes the laser-sensitive equipment and scores a hit. The TES requires Soldiers to aim accurately and to use cover and concealment effectively. When used with an OPFOR, TES offers realistic training. It is an excellent training aid for an FTX or STX. It is also effective at the shoot house as a substitute for live fire. The two major shortcomings of TES are that it allows Soldiers to seek cover behind structures that normally would not protect them, and that it does not replicate suppression.

NAVIGATION

A-19. Navigational techniques for urban areas differ from those needed on other terrain.

- A typical 1:50,000-scale military map provides insufficient detail for accurate navigation. Therefore, Soldiers must learn to navigate using commercial maps, sketches, and aerial photographs. Leaders must learn how to plan operations using these items.
- Underground navigation is a special challenge. Soldiers must know subway plans and be able to identify tunnels, elevated tracks, stairways, and other critical features. They must learn how to measure distance, remain oriented, and mark their routes.
- Destruction often erases urban features that could aid in navigation. Soldiers must learn to orient using distinctive, difficult-to-erase features such as bridges, waterways, towers, and intersections.
- Leaders conduct navigational exercises in an actual urban area. Soldiers must move from point to point while controllers at each point monitor and verify Soldiers' progress.
- The GPS is a navigational aid that triangulates with satellites to calculate position. However, this system may be affected by nearby tall buildings, and its effectiveness decreases inside buildings or underground.

LIMITED VISIBILITY TRAINING

A-20. Urban operations involve day and night operations. FTXs, STXs, CFXs, and CPXs are the most valuable exercises for night training.

- Field-expedient lighting, NVDs, illumination, and electrical lighting must be exploited when the tactical situation allows. Soldiers must learn discipline in their use of light, and also learn the lighting techniques that allow them to carry out their urban training program.

- Situational awareness is the Soldier's knowledge of his location, the locations of friendly forces, and the locations of the enemy on the battlefield. Situational awareness is the key to avoiding fratricide, especially during limited visibility operations. Units must plan to use all the equipment and devices available to mark friendly units and to avoid fratricide, without exposing themselves to enemy observation.
- Reverse-cycle training should be part of an urban training program. Limited visibility fighting skills are perishable, but they can be sustained and improved by effective training. The goal is for every Soldier to operate as effectively at night as during the day.

SAFETY AND AMMUNITION CONSIDERATIONS

A-21. When using ammunition and pyrotechnics, all associated safety requirements must be followed. This paragraph describes some of the safety equipment and training procedures to consider:

SPECIAL EFFECT SMALL-ARMS MARKING SYSTEM

A-22. SESAMS is 5.56-mm and 9-mm color-marking training ammunition. It is designed for close combat training. It is used with the close combat mission capability kit (CCMCK) adapter kit, which is installed on the M4/16 rifle, M9/M11 pistol, and M249 SAW. Different colors can be assigned to OPFORs for engagement identification and fratricide tracking. SESAMS has a 30-meter effective range and activates TES (MILES), so the two are used together. Anytime during a SESAMS exercise that a Soldier loses any of his protective gear, he must immediately cover his face and eyes with his hands. He must stay put until the engagement ends, or until he is told that it is safe to replace his safety equipment. Everyone, to include observer-controllers, must wear the protection shown in Table A-2.

Table A-2. SESAMS protection.

SKIN	Wear gloves, and cover all exposed skin. At close ranges, the SESAMS projectile can break skin.
HEAD	Take every precaution to protect your head.
EYES	Keep your eyes completely enclosed within eye protection. You can use standard Army sun, wind, and dust goggles.
EARS	Wear hearing protection to prevent hearing loss.
MOUTH	Keep your mouth covered.

DANGER

Failure to wear the required protection during SESAMS exercises can result in injury or death.

M84 STUN GRENADE

A-23. The hand-deployed M84 stun grenade serves as a reliable, effective, nonlethal means of neutralizing, diverting, and disorienting enemy personnel. It is used tactically as well as in training.

A-24. The M84 produces an intense (1 to 2.5-million candlepower) flash and, at 1.5 meters, a 170 dBs bang. Even though this is too low to produce permanent hearing damage, one-time use Soldiers are required to wear hearing protection anytime noise levels exceed 85 dBs.

A-25. The M84 stun grenade has an M201A1 fuze and a pyrotechnic output charge in a synthetic casing with a one- to two-second delay fuze, which cannot be cooked off. The grenade body has a nonfragmenting metal hexagonal casing, which, once deployed, resists rolling.

SMOKE GRENADES

A-26. Smoke grenades are used in urban combat to cover movement, deceive enemy forces as to the exact location of the attacking force, and for signaling. Smoke grenades should not be used in buildings and subterranean passages as they displace oxygen and protective masks do not filter the smoke or provide oxygen. Buildings in urban areas channel air currents, which influence emplacement of smoke grenades. One grenade should be used to test the airflow before employing additional grenades.

WARNING

Avoid using smoke grenades in buildings and in subterranean passages.

OPPOSING FORCES

A-27. OPFOR should be used as often as possible to enhance the realism of training, especially for FTXs, CFXs, and STXs. Thebe realistic, the OPFOR must have defined countertasks. These countertasks must support "Blue Force" training, and should prevent accidental contact between the Blue Force and the OPFOR. Leaders should develop a thorough control plan for OPFOR scenarios. This plan should describe uniform and vehicle markings, ammunition, safety instructions, controller guidance, and guidelines for handling enemy prisoners of war (EPW) and noncombatants. The ROE must clearly specify the limits of close combat between Soldiers.

- OPFOR heighten Soldiers' interest by giving them a real opponent. This fosters unit competition and realism, especially when the OPFOR is used in addition to TES and SESAM.
- OPFOR challenges leaders' ability to improvise. However, this free-play scenario should be monitored to ensure the unit meets the training objectives.
- OPFOR can use threat urban operations and doctrine to determine their course of action. If available, OPFOR uniforms, mock documents, overlays, and equipment add realism and intelligence to the scenario. Leaders conduct preliminary training for the OPFOR, which helps the OPFOR recreate the threat's urban tactics.
- Noncombatants should be integrated into training scenarios.

PRECISION MARKSMANSHIP TRAINING

A-28. Precision marksmanship training should be conducted before training at the UAC, CACTF, and the Shoot House, which accommodate training in target discrimination (FM 3-22.9). This is the best use of the unit's time and resources.

PHYSICAL READINESS TRAINING

A-29. Combat in urban areas places extraordinary physical demands on Soldiers. Therefore, physical readiness training, with emphasis on upper body development, is paramount to any urban operations training program.

VARIED SCENARIOS

A-30. Leaders must vary training scenarios to prevent redundant and predictable tactical behavior. Using predictable avenues of approach into the facility produces stereotypical tactical solutions.

BUILDING FACADE WALLS, WINDOWS, AND DOORS

A-31. The use of façade walls, windows and doors, in conjunction with SESAMS in the interior of some stations at the UAC, provides for more realistic training. This method of training reinforces the fact that walls, windows, and doors in most cases *do not* stop the penetration of service ammunition. Plastic wrap sheeting, tar paper, or some other type of material can be used to allow SESAMS penetration (Figure A-1).

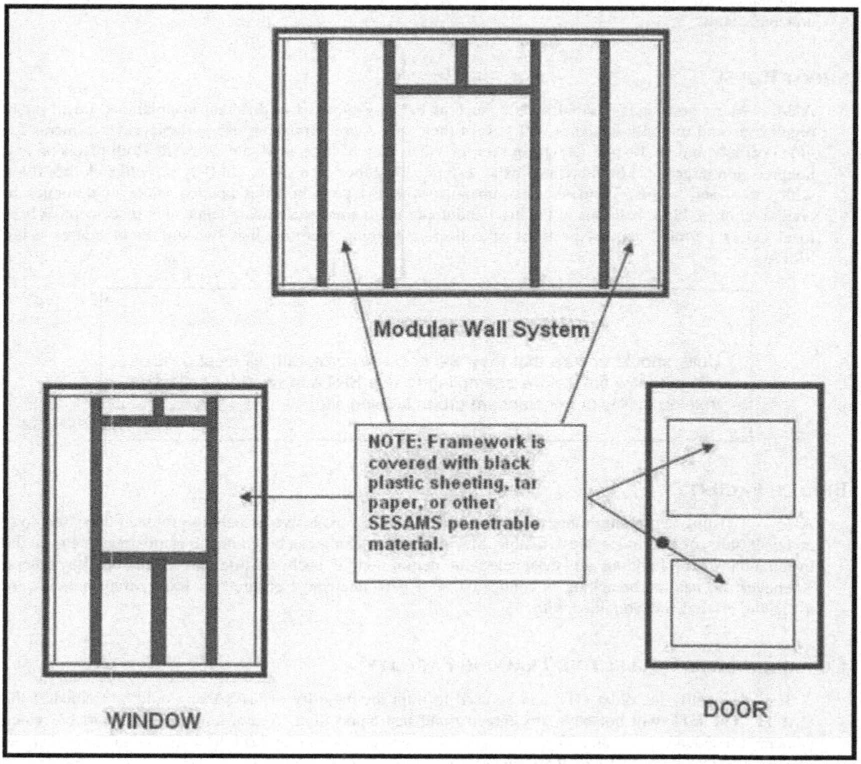

Figure A-1. Shoot-through walls and doors.

USE OF OLDER OR NONSTANDARD FACILITIES

A-32. Until installations receive funding for the UAC, SH, BF, and CACTF, they can use the older MOUT Assault Course (MAC) and Collective Training Facility (CTF) as described in the 2002 edition of TC 90-1, which this edition supersedes. Installations may also use nonstandard shoot houses and other nonstandard urban facilities. These facilities may be adapted as follows:

URBAN ASSAULT COURSE

A-33. Generally, the MAC can be used to train most of the UAC tasks. Of course, the MAC lacks targetry and instrumentation. Instead, units can record training with portable video cameras. During the AARs, they can play back the videos on VCRs hooked to TVs or on TV/VCR combos. OPFOR combatants will have SESAMS or MILES. People in civilian clothes can play the parts of the noncombatants.

SHOOT HOUSE

A-34. Many nonstandard shoot houses, such as tire houses, exist in different installations. Local range regulations and unit risk analyses will govern their use. Again, units can use portable video cameras for video capture and VCRs and television sets for video play back to facilitate an AAR. Both precision and nonprecision targets can be fabricated using E-type silhouettes on a stake, old BDUs/civilian clothes filled with straw, and so on. Simulate noncombatants with E-type silhouettes painted white or dummies in civilian clothes. Place balloons in the head and upper torso areas to simulate lethal shot placement. Where local policy permits, convert portions of collective training facilities into live-fire shoot houses using SRTA.

CAUTION

Units should ensure that they are in compliance with all local policies affecting live fire before attempting to use SRTA in an older collective training facility or nonstandard urban training site.

BREACH FACILITY

A-35. Existing demolition ranges may be used to conduct explosive breaching provided door, window, and wall units, or facsimiles, are available. Mechanical breaching can be trained in cantonment areas on the installation where building are designated for demolition, if such facilities are available. Regardless, whenever mechanical breacking is conducted, unit personnel must ensure that local safety policies are available, briefed, and complied with.

COMBINED ARMS COLLECTIVE TRAINING FACILITY

A-36. Generally the older CTF can be used to train the majority of tasks that would be trained in the CACTF. The CTF will not have any targetry and instrumentation. Again, units can use portable video cameras for video capture and VCRs and television sets for video play back to facilitate an AAR. Targets can be OPFOR with MILES. Role players in civilian clothes can be introduced to play the part of noncombatants. If armored vehicles are used, units should ensure that local range policies allow for their use. Older CTFs are normally either a 16 or 32 building design. Appropriate unit-level training, platoon, company, or battalion, should be planned based on the size of the facility.

Appendix B

Unit Training Matrix

Unit training plans are prepared IAW FM 7-1, which states that "Planning links the unit METL and the execution of battle-focused training.". The commander begins his planning with guidance from a higher echelon and an in-depth assessment of METL tasks (in this case, mission-essential urban tasks). Based on his own observations, feedback from his subordinates, and other sources the commander determines which tasks should be trained, which should be trained first, and how much time should be allotted for each. Tables B-1 through B-5 list training tasks by the WFF to be trained at each facility.

Table B-1. Army unit training matrix, brigade tasks.

Brigade Tasks	***SHOOT HOUSE*** *Semiannual*	***BREACH HOUSE*** *Semiannual*	***UAC*** *Quarterly*	***CA/CTF*** *Semiannual*	***CTC*** **As Scheduled*	***SIMULATION*** *Annual*
Develop Intelligence						
Perform Intelligence Operations					X	X
Perform S-2 Operations					X	X
Deploy/Conduct Maneuver						
Attack					X	X
Perform Air Assault					X	X
Defend					X	X
Lodgment					X	X
Exercise Command and Control						
Command and Control the Brigade					X	X
Maintain Communications					X	X
Coordinate Air Defense Operation					X	X
Conduct Noncombatant Evacuation					X	X
Develop a Media Plan					X	
Conduct Mediation and Negotiation					X	
Employ Firepower						
Execute Fires					X	X
Employ Air Support					X	X
Employ Attack Helicopters					X	X
Synchronize Fire Support					X	X
Synchronize CAS					X	X
Conduct Targeting Process					X	X
Protect the Force						
Perform Mobility/Survivability Operations					X	X
Provide Engineer Support					X	X
Coordinate CBRN Operations					X	X
Conduct Operations Security					X	X
Perform Sustainment						
Evacuate Casualties					X	X
Treat Casualties					X	X
Sustain the Force					X	X
Conduct a Civil-Military Operation					X	X
Provide Humanitarian Support					X	X

Table B-2. Army unit training matrix, battalion tasks.

Battalion Tasks	*SHOOT HOUSE Semiannual*	*BREACH HOUSE Semiannual*	*UAC Quarterly*	*CA/CTF Semiannual*	*CTC As Scheduled*	*SIMULATION Annual*
Develop Intelligence						
Conduct S-2 Operations				X	X	X
Deploy/Conduct Maneuver						
Attack a Built-Up Area				X	X	X
Plan for Urban Operation				X	X	X
Defend an Urban Area				X	X	X
Conduct an Infiltration/Exfiltration				X	X	X
Perform Cordon and Search Operations in an Urban Area				X	X	X
Plan for Urban Operations				X	X	X
Conduct Tactical Movement				X	X	X
Conduct Presence Operation in a Stability Environment				X	X	X
Defend an Urban Area				X	X	X
Traverse a Built-up Area				X	X	X
Conduct a Mobile Defense				X	X	X
Employ Firepower						
Integrate Fires and Effects Support				X	X	X
Establish the Fire Support Element				X	X	X
Protect the Force						
Conduct Mobility or Survivability Operations				X	X	X
Secure Civilians During Operation				X	X	X
Conduct Passive Air Defense Measures				X	X	X
Conduct Active Air Defense Measures				X	X	X
Employ Operations Security Measures				X	X	X
Integrate Air Defense Support				X	X	X
Perform Sustainment						
Conduct S-1 Operations				X	X	X
Conduct S-4 Operations				X	X	X
Conduct a Civil-Military Operation				X	X	X
Exercise Command and Control						
Conduct S-6 Operations				X	X	X
Conduct Noncombatant Evacuation				X	X	X
Conduct S-3 Operations				X	X	X
Conduct Mediation and Negotiations				X	X	X
Plan Roadblocks or Checkpoints				X	X	X
Establish Liaison				X	X	X

Table B-3. Army unit training matrix, company tasks.

Company Tasks	*SHOOT HOUSE* *Semiannual*	*BREACH HOUSE* *Semiannual*	*UAC* *Quarterly*	*CA/CTF* *Semiannual*
Develop Intelligence				
Conduct an Area or Zone Reconnaissance				X
Reconnoiter a Built-Up Area				X
Deploy/Conduct Maneuver				
Conduct Cordon and Search				
in a Built-Up Area				X
Attack a Built-Up Area				X
Conduct Presence Operations				X
Conduct Tactical Movement				
in a Built-Up Area				X
Defend in Urban Environment				X
Conduct an Infiltration (Co/Plt)				X
Conduct Operations with Armored or				
Mechanized Infantry Vehicles				
in an Urban Environment				X
Conduct Civil Disturbance Operations				X
Establish Checkpoints				X
Employ Firepower				
Integrate Direct Fires				X
Integrate Indirect Fires				X
Protect the Force				
Maintain Operations Security				X
Conduct Passive Air Defense Measures				X
Conduct Passive Air Defense Measures				X
Perform Sustainment				
Treat Casualties				X
Evacuate Casualties				X
Secure Civilians during Operations				X
Conduct Resupply Operations				
Process Captured Documents and				X
Equipment				X
Handle Enemy Prisoners of War				
Exercise Command and Control				
Prepare for Combat				X
Establish Communications				X
Conduct Consolidation and Reorganization				X
Conduct Negotiation				X
Conduct Troop-Leading Procedures				X

Table B-4. Army unit training matrix, platoon/squad tasks.

Platoon/Squad Tasks	*SHOOT HOUSE Semiannual*	*BREACH HOUSE Semiannual*	*UAC Quarterly*	*CA/CTF Semiannual*
Develop Intelligence				
Report Tactical Information	X		X	X
Conduct an Area or Zone Reconnaissance	X		X	X
Reconnoiter a Built-Up Area	X		X	X
Deploy/Conduct Maneuver				
Conduct Tactical Movement in a Built-Up Area	X		X	X
Assault a Building	X		X	X
Conduct a Strongpoint Defense of a Building				X
Conduct an Infiltration or Exfiltration				X
Conduct Operations with Armored or Mechanized Vehicles in an Urban Environment	X			X
Conduct a Presence Patrol	X			X
React to a Civil Disturbance				X
Establish a Checkpoint				X
Search a Building	X		X	X
Employ Firepower				
Employ Fire Support				X
Protect the Force				
Maintain Operations Security				X
Conduct Passive Air Defense Measures				X
Conduct Active Air Defense Measures				X
Perform Sustainment				
Secure Civilians During Operations	X		X	X
Perform Resupply Operations				X
Treat Casualties	X		X	X
Evacuate Casualties	X		X	X
Handle Enemy Prisoners of War				
Exercise Command and Control				
Conduct Troop-Leading Procedures	X		X	X
Establish Radio Communication	X		X	X
Conduct Consolidation and Reorganization	X		X	X
Prepare for Combat	X		X	X

Table B-5. Army unit training matrix, individual tasks/techniques.

Individual Tasks/Techniques	*SHOOT HOUSE Semiannual*	*BREACH HOUSE Semiannual*	*UAC Quarterly*	*CA/CTF Semiannual*
Skill Level 1				
Perform Movement Techniques in an Urban Environment, 071-326-0541	X	X	X	X
Prepare Positions for Individual and Crew-Served Weapons in an Urban Environment, 071-326-0550	X	X	X	X
Select Hasty Fighting Positions in an Urban Environment, 071-326-0557	X	X	X	X
Employ Hand Grenades, 071-325-4407	X	X	X	X
Engage Targets with an M16-Series or M16A2 Rifle, 071-311-2007	X	X	X	X
Execute a Mechanical Breach, 071-440-0027			X	X
Skill Level 2				
Conduct an Explosive Breach	X	X	X	X
Conduct a Breach, 071-440-0029	X	X	X	X
Prime Explosive Nonelectrically, 051-193-1003	X	X	X	X
Construct a Nonelectric Initiating Assembly with Modernized Demolition	X	X	X	X

Appendix C

Stability and Support Operations Unit Training Matrix

When assigned a stability or support mission, a well-trained unit must be able to rapidly shift its focus from war fighting to stability and support and also from stability and support to war fighting. During a stability or support operation, the unit performs numerous activities. Essentially, the unit accomplishes these activities through execution of tactical missions and tasks. While stability and support operations can occur anywhere, it is most probable that they will occur in an urban environment. Tables C-1 through C-4 list training tasks by the WFF to be trained at each facility.

Table C-1. Stability and support operations unit training matrix, brigade tasks.

Brigade Tasks	SHOOT HOUSE	BREACH HOUSE	UAC	CA/CTF	CTC	SIMULATION
Develop Intelligence						
Collect and Disseminate Information					X	X
Deploy/Conduct Maneuver						
Demonstrate a Show of Force					X	
Employ Psychological Operation					X	X
Employ Firepower						
Provide Fire Support					X	X
Protect the Force						
Protect the Force					X	
Perform Sustainment						
Restore Law and Order					X	
Exercise Command and Control						
Establish an Evacuation Control Center/Conduct Evacuation Control Center Operations					X	
Negotiate					X	
Perform Noncombatant Evacuation Operations					X	
Plan a Media Visit					X	
Develop Intelligence						
Collect and Disseminate Information				X	X	X

Table C-2. Stability and support operations unit training matrix, battalion tasks.

Battalion Tasks	*SHOOT HOUSE*	*BREACH HOUSE*	*UAC*	*CA/CTF*	*CTC*	*SIMULATION*
Deploy/Conduct Maneuver						
Interdict Smuggling Operations				X	X	
Disarm Belligerents				X	X	
Handle Captured Insurgents or Belligerents				X	X	
Employ Psychological Operation				X	X	
React to Civil Disturbance				X	X	
Employ Firepower						
Provide Fire Support				X	X	X
Protect the Force						
Protect the Force				X	X	
Restore Law and Order						
Identify and Process Detainees				X	X	
Deliver Supplies or Humanitarian Aid				X	X	
Control Civilian Movement				X	X	
Prepare Traffic Control Plan				X	X	
Exercise Command and Control						
Negotiate				X	X	
Perform Noncombatant Evacuation Operations				X	X	
Conduct Marshalling Force Operations				X	X	
Negotiate a Belligerent Force Checkpoint				X	X	
Plan a Media Visit				X	X	
Perform a Cordon and Search				X	X	
Coordinate and Monitor Convoy Security				X	X	

Table C-3. Stability and support operations unit training matrix, company tasks.

Company Tasks	*SHOOT HOUSE*	*BREACH HOUSE*	*UAC*	*CA/CTF*	*CTC*	*SIMULATION*
Develop Intelligence						
Collect and Disseminate Information				X	X	
Deploy/Conduct Maneuver						
Interdict Smuggling Operations				X	X	
Disarm Belligerents				X	X	
Handle Captured Insurgents or Belligerents				X	X	
Defend Convoy				X	X	
Apprehend/Detain Noncombatants				X	X	
Perform Cordon and Search				X	X	
Employ Firepower						
Provide Fire Support				X	X	
Protect the Force						
Establish a Checkpoint				X	X	
Operate a Checkpoint				X	X	
Protect the Force				X	X	
Perform Sustainment						
Process Captured Documents and Equipment				X	X	
Exercise Command and Control						
Negotiate a Belligerent Force Checkpoint				X	X	
Conduct Marshaling Force Operations				X	X	
Negotiate a Belligerent Force Checkpoint				X	X	
Plan a Media Visit				X	X	
Coordinate and Monitor Convoy Security				X	X	
Link Up with a Convoy				X	X	

Table C-4. Stability and support operations unit training matrix, platoon/squad tasks.

Platoon/Squad Tasks	*SHOOT HOUSE*	*BREACH HOUSE*	*UAC*	*CA/CTF*	*CTC*	*SIMULATION*
Deploy/Conduct Maneuver						
Disarm Belligerents				X	X	
Handle Captured Insurgents or Belligerents				X	X	
Apprehend/Detain Noncombatants				X	X	
Search a Building	X			X	X	
React to a Sniper			X	X	X	
Conduct Platoon Riot Control Formation				X	X	
Protect the Force						
Establish a Checkpoint				X	X	
Operate a Checkpoint				X	X	
Protect the Force				X	X	
Exercise Command and Control						
Negotiate a Belligerent Force Checkpoint				X	X	
Exercise Command and Control						
React to a Media Interview				X	X	

Appendix D

Leader Development Training

To be successful in combat, the Army must continually train to develop and maintain combat-ready Soldiers, leaders, and units that can perform assigned tasks to specific standards.

SIMULATIONS TRAINING

D-1. Commanders train their staffs to plan, coordinate, and synchronize their units' operations from initial mobilization and deployment. These operations include the full spectrum of Army operations; offense, defense, stability, and support. Current simulations lack adequate replication of urban scenarios, terrain, and weapons effects. The introduction of WARSIM 2000 and OneSAF should help to solve these simulations modeling deficiencies. For lower echelon units, battalion and below, constructive simulations complement collective battle staff training that occurs during virtual simulation similar to the close combat tactical trainer and live events. For echelons above battalion, it becomes difficult to use simulators or live maneuver to train. Studies to date indicate the crossover point for simulation versus simulator is at the battalion level. Some training events from battalion and below can be trained effectively using various types of simulators. Above battalion level, effective training almost has to be conducted with a simulation (Figure D-1).

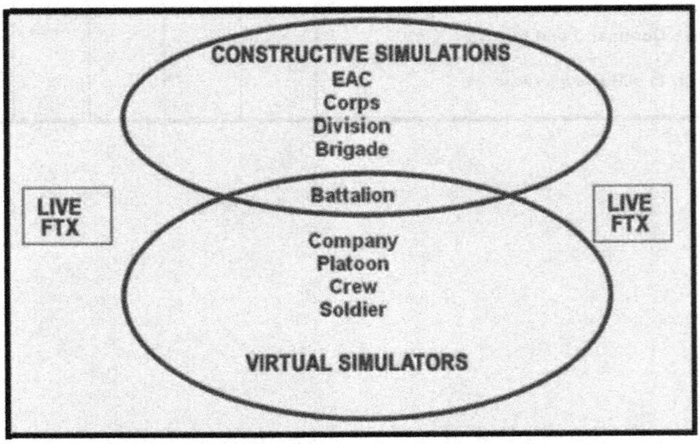

Figure D-1. Echelons and uses.

D-2. Simulations that support leader and battle staff training focus on collective tasks rather than individual leader tasks. Simulations, used in this context, rely on the human dynamics in the command setting and use real-time actions in a tactical environment with a unit's go-to-war systems. This process forces commander and staff interaction that provides feedback and helps refine the integration process. Battle staff training encourages team building, task coordination, and standardization of staff actions.

- Constructive simulations can provide commanders large areas for conducting training, access to sensitive areas through computer and mapping technology, and futuristically provide a much better level of fidelity in terms of urban operations terrain and scenario replication.
- Simulations can portray large, capable, and doctrinally correct OPFORs. All of the Soldiers portrayed in the units in a simulation do not have to be present in order for the training to be realistic. An attractive feature of simulations has always been the reduced staffing necessary to conduct this type of training exercise.
- Simulations can stress CPs and provide the realistic conditions under which commanders must make decisions. Brigade and below simulation (BBS) and corp battle simulation (CBS) are well known for the stressful environments created during an exercise. The stress is realistic and can assist in conditioning the training audience before CTC rotations.
- Simulations allow different units to train under the same conditions and to the same standards. This technique is a unique feature of simulations as the exact weather, opposing force organization, location of units, and equipment can be used to train many different units on the same terrain with the same conditions.
- The size and number of personnel in a battle staff vary depending on the echelon and other demands unique to that commander. However, the WFF are represented by the following staff officers: intelligence, operations, fire support, communications, logistics, air defense, and engineer.
- Characteristics and benefits of simulations used to support battle staff training follow:
  - -- Build and sustain staff teamwork.
  - -- Build and sustain effective and efficient organizations.
  - -- Provide training and validation of command and staff procedures (SOPs, TACSOPs, FSOPs).
  - -- Provide proficiency to the individual staff officer, while building proficiency in a fully integrated staff.
  - -- Allow practice of battle drills and tactical missions (movement to contact, hasty defense).
  - -- Allow commanders to determine if their staff and subordinate commanders understand the commander's intent and concept of operations.

D-3. The major development efforts for FAMSIM is WARSIM 2000. WARSIM 2000, in conjunction with OneSAF, exploits new technologies to enable commanders and staffs at battalion through EAC to train in a realistic simulation environment (Table D-1). The increased realism of WARSIM 2000 over existing models allows units to use Army Battle Command Systems (ABCS) to synchronize across the complete range of operations. WARSIM 2000's design allows warfighting CPs to use organic ABCS equipment to interact with the simulation from training sites in the field, as well as from simulation centers. WARSIM 2000 will also be capable of showing a joint and combined environment across the operational continuum.

Table D-1. Urban operations simulations crosswalk.

CAMTF Training Tasks	*WARSIM*	*OneSAF*
Corps	X	
Division	X	
Brigade		X
Battalion		X
Company		X
Platoon		X

D-4. Future simulation capabilities simultaneously integrate live, virtual, and constructive environments. Constructive simulation is a viable solution for battalion and above training and the federation of WARSIM and OneSAF models should alleviate the following training simulation problems:

- Eliminate the need for swivel chair interface.
- Allows focus on subordinate units fighting the fight.
- Digital and analog feeds to C4I systems including FBCB2.
- May provide for distance learning opportunities.
- Facilitates simultaneous multi-echelon urban operations training

D-5. At company and below, the best training solution is live and virtual. This is accomplished by linking the Close Combat Tactical Trainer (CCTT) to OneSAF and VLET to Engagement Skills Trainer (EST) technology. Entity-based models like OneSAF can allow the leader to fight from a TOC or vehicle, if linked to virtual or live environments. It also has the potential for distance learning.

PHASES OF TRAINING

D-6. The Army Urban Operations Training Strategy consists of initial, intermediate, and culminating stages. These phases of training, shown in Figure D-2, describe the sequence in which urban operations training tasks should be conducted.

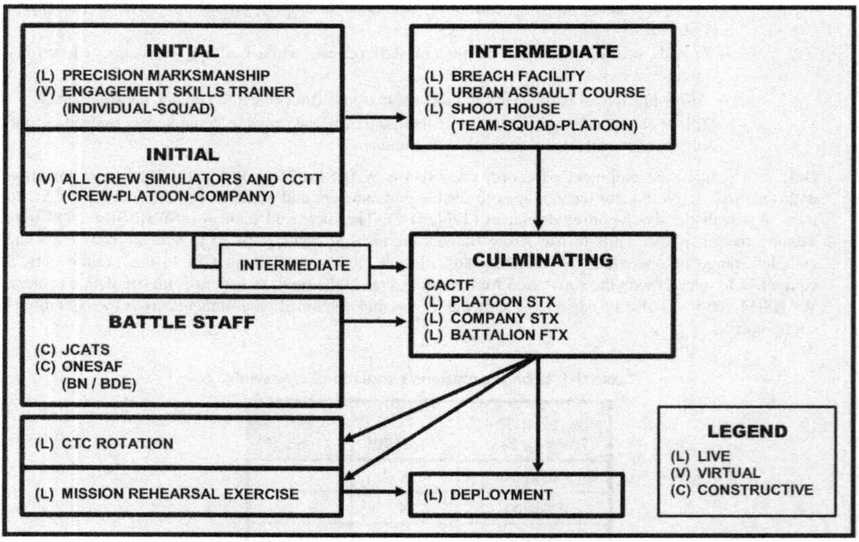

Figure D-2. Phases in the Live, Virtual, and Constructive Training Strategy.

Initial Phase

D-7. The initial phase of training covers those individual and collective tasks that Soldiers and leaders should be proficient at before arriving at any of the urban operations training sites. They consist primarily of basic marksmanship skills and Soldier tasks needed to perform any tactical operation.

Intermediate Phase

D-8. The intermediate phase consists of training conducted at the breach facility, shoot house, and the urban assault course. Urban operations training tasks should be conducted here before advancing to the Combined Arms Collective Training Facility with the exception of Aviation, Armor, and other branches that have no practical use for the Shoot House, Breach Facility, or Urban Assault Course.

Culminating Phase

D-9. The combined arms collective training facility is the culminating phase of urban operations training and focuses on multiechelon collective skills for all branches of the combined arms team. Urban operations leader training currently exists in our leader professional development courses. Units have developed internal urban warfare training programs and TRADOC schools have assembled technical mobile training teams (MTTs) for marksmanship, combatives, and breaching techniques.

Education Phase

- Urban warfare reading/video list.
- Urban operations equipment.
- Review of doctrinal publications (FM 3-21.10, FM 3-06.11, for Infantry) and corresponding publications for other branches.
- Small-unit urban movement techniques and considerations.

Planning and Execution Phase

- Prepare company OPORD or WFF-related products.
- Brief platoon/company OPORD/WFF-related products to commander or staff representative.
- Participate as briefer in urban environment TEWT.

Sustainment Phase

- Participate in company/platoon TEWT or other urban operations event.
- Execute/facilitate a company or platoon TEWT.

EXAMPLE LEADER'S COURSE POI

D-10. An example urban leaders' course POI, which is intended to provide unit trainers with a tailorable urban operations training program, is shown in Table D-2. It is a *train-the-trainer* course for team leaders through platoon leaders to train tactics, techniques, and procedures of advanced urban operations skills. The length and focus of the POI can be tailored to support the commander's training objectives and unit METL. The target audience is newly arrived, or soon to be, small unit leaders. The recommended course length is about two weeks or 120 to 140 hours of instruction.

REALISTIC TRAINING OFF FEDERAL FACILITIES

D-11. Realistic training for urban operations is critical to force readiness. Training opportunities that are both realistic and challenging may be unavailable at federal installations, or at facilities specifically designed for urban operations training. Training in off-post civilian urban settings may become necessary in order to support readiness and training requirements.

D-12. The Department of Defense has developed the following SOP for planning and approving such events. The SOP assists in balancing the needs of the military training and of the civilian community/ies to be used for realistic urban training events.

Table D-2. POI for urban leader's train-the-trainer course.

Course Overview	• Introduction	
Precision Marksmanship	• Firing techniques. • Target discrimination.	
Breaching Techniques	• Explosive. • Ballistic. • Mechanical.	

Table D-2. POI for urban leader's train-the-trainer course (continued).

Offensive Operations	• Planning considerations	-- METT-TC. -- Snipers. -- CASEVAC. -- Communications. -- ROE.
	• Platoon attack a built-up area (task organization).......................................	-- Assault (isolate, enter, clear). -- Rooms, hallway, stairs, superstructure, subsurface. -- Markings. -- Movement techniques. -- Streets. -- Alley ways. -- Obstacles. -- Other danger areas. -- Open areas. -- Roof tops.
	• Combined arms	-- Armor/mechanized platforms. -- Aviation assets. -- Artillery. -- Engineer.
	• Limited visibility	
Defensive Operations	• Planning considerations	-- METT-TC. -- Snipers. -- CASEVAC. -- Communications. -- ROE.
	• Characteristics of urban areas	
	• Hasty defense	-- Occupation and preparation of positions. -- Hasty firing positions. -- Improving the defense.
	• Deliberate defense, including priorities of work and other considerations	
	• Defensive plan at platoon level.	-- Defense of a strongpoint. -- Defense against armor.
	• Combined arms	-- Armor/mechanized. -- Aviation assets. -- Artillery. -- Engineer.
	• Limited visibility • Fundamental combat skills	

Table D-2. POI for urban leader's train-the-trainer course (continued).

Stability and Support Operations	• Principles of stability and support.	
	• Stability and support activities	-- NEO. -- Humanitarian assistance. -- Disaster relief. -- Show of force. -- Attack and raids.
	• Rules of engagement application • Convoy operations • Checkpoints • Fire support • Sniper employment/counter sniper measures • Civil disturbance	
Urban Training Facilities	• Breach facility • Urban assault course • Shoot house • CACTF	
New Technologies	• Rifle launch entry munition	
	• Unmanned vehicles	-- UAV. -- UGV. -- Ladders. -- Night vision devices.

STANDING OPERATING PROCEDURES FOR
PLANNING AND APPROVING REALISTIC URBAN TRAINING

1. **REFERENCES.**

    10 USC § 167, *Unified Combatant Command for Special Operations Forces*.

    10 USC §§ 3013, 5013, 8013.

    Department of Defense Directive 5111.10.

    Unified Command Plan 99.

2. **PURPOSE**. This SOP establishes uniform planning and approval procedures for realistic urban training (RUT) events.

3. **APPLICABILITY**. This SOP applies to all RUT events in civilian urban settings within the United States, its territories, and its possessions; and to Active or Reserve Forces, including National Guard Forces in Federal service. This policy does not apply to National Guard Forces under in-state status (Title 32), or to aviation operations that do not require an exemption from standard Federal Aviation Regulations. The training event should not be located on federal state, local, or private installations or facilities.

4. **DEFINITIONS**. For the purposes of this SOP, the following definitions apply:

a. **Urban Operations. This term** includes (see FM 3-06) all military actions planned and conducted on a topographical complex, and on adjacent natural terrain, where man-made construction is the dominant feature. It includes combat in towns and cities, including house-to-house and street-by-street fighting.

b. **Realistic Urban Training.** RUT is high-intensity, close-quarter battle training using live or nonlethal fires; demolitions or explosives such as breaching charges; and air and naval supporting platforms at the objective, in ***civilian urban settings***.

c. **Civilian Urban Setting.** A civilian urban setting is any area(s) located in or near a civilian community. Civilian urban settings will not be developed or established for training events, especially where training activities would affect civilian residences or commercial areas due to noise, traffic, and so on.

5. **PROCEDURES.** The following procedures apply to the planning and approval of RUT in civilian urban settings:

a. The combatant commander or the service ensures compliance with this SOP. Approval authority for RUT events may be delegated to a general or flag officer within the chain of command, or to the combatant common service conducting the training.

b. The approval authority reviews proposed training events to determine whether they are necessary and appropriate for unit readiness. Training must support unit METs and JMETs.

c. On approval of the proposed training, and at least 14 days before training starts, the combatant common service notifies the **SECR/PF** through the CJCS (by way of message) of the approved training. At a minimum, this notice includes the following:

- Name and location of the unit conducting training.
- Scope of training and activities planned such as live fire, breaches, or use of aircraft.
- Start and end dates.
- Training location.
- A statement that all necessary coordination has been accomplished IAW this SOP. The statement should include a list of the contacted official's names and their positions.

d. **Coordination with Civilian Officials**.

- While planning the training event, the commander coordinates with and obtains written approval to conduct the event from the appropriate federal, state, and local civilian authorities. Commanders will consult their legal and public affairs officers to identify the civilian officials, for example the mayor, borough chief, county commissioners or supervisors, with whom they must coordinate the training event. To do this, they consult local and Federal law enforcement agencies. At a minimum, they must consult a senior official with responsibility for each affected civilian urban setting.
- During coordination, the commander covers the details of the proposed training activities, such as whether the event involves live fire, rotary- or fixed-wing aircraft, or close quarters battle training; and whether the exercises will be conducted during the day, night, or both: and so on. It will also list actions to ensure public and participant safety, proposed public or resident notification and outreach procedures, and any other issues that might cause public interest or concern.
- As much as possible, the commander briefs civilian officials in person on the exercise concept.

- The US congressional representative or staff for the training district, and the US senators for the state, or their staffs, are advised of the proposed training event and given whatever information they request. Neither written approval nor concurrence from the representative or senators is required. However, if these officials raise objections or issues, and if those issues cannot be resolved, they are immediately elevated to the Joint Staff, J3, and the service or combatant commander point of contact.
- For each training site, license agreements with civilian or government property owners or managers are executed. Agreements address liability and conditions of acceptable usage of property. Property owners are informed of the nature of the training including written notification as to whether the activities will include use of breaching charges and, if so, how much; or live ammunition and, if so, which weapons and what ammunition. This notice will also state *if* rotary or fixed-wing aircraft will be used *and* whether the exercises will be conducted during the day, night, or both. Before the exercise can begin, all license agreements must be reviewed for legality. Copies of the license agreements will be provided to the senior civilian officials responsible for training area.
- A record of coordination activities is maintained for each training event. The record includes the names and positions of the officials *who* coordinate and approve the event. It also provides officials with a summary of the events such as date, time, and location of the training; types of activities; and planned public notification and outreach. A follow-on letter is distributed to the senior civilian officials responsible for the affected civilian urban settings. The letter confirms the training event *and* lists the names of the civilian officials who plan the training event.
- The unit may schedule recurring training events at a particular area or location by submitting a single MOA to the authorizing civilian officials. The terms of the MOA must comply with the requirements of this SOP. The service or the combatant commander reviews the MOA, which authorizes the recurring training, to ensure it complies with the procedures. Once the commander approves the MOA, the unit may conduct individual training events under it without further Joint Staff or OSD review. However, services and combatant commanders notify SO/LIC *and* the Joint Staff 14 days *before* the training in this MOA begins. Thereafter, services and combatant commanders must review these MOAs periodically to ensure *that* they continue to fulfil the intent.
- *If* nondisclosure agreements are required, command legal representatives must review them. Nondisclosure statements should clearly explain *that* the agreement does not preclude persons from informing others of general, nonsensitive information about the training exercise.
- Depending on OPSEC and safety considerations, commanders must adhere to the recommendations of civilian officials regarding community notification, outreach, and press procedures. Community notification and outreach procedures may include door-to-door notification by the local law enforcement agency. Commanders will advise civilian officials when they cannot follow their recommendations.

e. Other instructions include--

- Legal and PA representatives attend final planning meetings, and they remain on-site throughout all training activities.
- The on-site PAO handles all press queries after coordinating with the on-scene commander.
- All requested training must follow applicable statutes, law, DOD directives, other guidance, and implementing service directives.

Glossary

Section I. ACRONYMS AND ABBREVIATIONS

1SG first sergeant

A

AA assembly area
AAR after-action review/report
ABCS Army Battle Command System
ACU Army combat uniform
ADA air defense artillery
ANCOC advanced noncommissioned officer course
AO area of operation
AR Army Reserve
AT4 a shoulder-fired munition, formerly called a "light antiarmor weapon"
atk attack
ATWESS antitank weapon effect signature simulator
AVCATT armored vehicle combined arms tactical trainer

B

BBS brigade and below simulation
BCT brigade combat team
bde brigade
BDU (obsolete; *see* ACU)
BF Breach Facility
bn battalion
BNCOC basic noncommissioned officer course
BOLC Basic Officer Leadership Course
BOS (obsolete; *see* WFF)
BP battle position
BSA brigade support area
BSX battle simulation exercise

C

C2 command and control
CACTF Combined Arms Collective Training Facility
CASEVAC casualty evacuation
CBRN chemical, biological, radiological, nuclear
CBS corps battle simulation
CCMCK close combat mission capability kit
CCOC Captain's Career Officer Course
CCTT close combat tactical trainer
CCTV closed-circuit television
CFX command field exercise
CMTC Citizens Military Training Corps
co company
commo communications
CP command post
CPX command post exercise
CTC Combat Training Center
CTF collective training facility

D

DES demolition effects simulator
det cord detonation cord
div division
DLIC detachment left in contact
DS Digital Training Management System
DTMS direct support

E

EA engagement area
EAC echelons above corps
ENCATT engineer combined arms tactical trainer
engr engineer
EPW enemy prisoners of war
EST Engagement Skills Trainer

F

FAMSIM family of simulations
FO forward observer
FRAGO fragmentary order
FSCATT fire support combined arms tactical trainer
FSOP field standing operating procedures
FTX field training exercise

G

GREM	grenade rifle entry munition
GSR	ground surveillance radar

H

HC	herachloroethane zinc
HEAT	high-explosive antitank

I

IAW	in accordance with
Inf	Infantry

J

JANUS	Joint Army-Navy Uniform Simulation
JCATS	Joint Conflict and Tactical Simulation
JMET	joint mission-essential task
JRTC	Joint Readiness Training Center
JTF	joint task force
JTX	joint training exercise

L

LAW	light antiarmor weapon
LFX	live-fire exercise
LMG	light machine gun

M

MAC	MOUT assault course
MACOM	major command
MAPEX	map exercise
M/C/S	mobility/countermobility/ survivability
MET	mission-essential task
METL	mission-essential task list
METT-TC	mission, enemy, terrain, troops, time available, and civilians
MI	military intelligence
MILES	Multiple Integrated Laser-Engagement System
MOPP	mission-oriented protective posture
MPAT	multipurpose antitank (ammunition)
MPAT-OR	multipurpose antitank-obstacle reducing
MRE	meals ready to eat
MTT	mobile training team

N

NBC	(obsolete; *see* CBRN)
NEO	noncombatant evacuation operations
NLT	not later than
NTC	National Training Center

O

OneSAF	one semiautomated force (software used to create modeling and simulation programs with military applications)
op	operation
OPFOR	opposing force
OPORD	operation order
OPSEC	operations security

P

plt	platoon
POI	program of instruction
PPP	power projection platform
PSP	power support platform

R

ROE	rules of engagement
ROI	rules of interaction
RUT	realistic urban training

S

S-1	personnel staff officer (adjutant)
S-2	intelligence staff officer
S-3	operations staff officer
S-4	logistics staff officer
S-5	civil affairs officer
S-6	command, control, communications and computer operations (C4 operations) officer
SAAF	small-arms alignment fixture
SBCT	Stryker brigade combat team
SDZ	surface danger zone
SEP	system enhancement program: a version of the M1A2 machine gun with enhanced digital command and control and second-generation forward-looking infrared

SESAMS	Special Effects Small Arms Marking System
SH	Shoot House
SOI	signal operation instructions
SOP	standing operating procedure
sqd	squad
SRTA	short-range training ammunition
STAFFEX	staff exercise
STX	situational training exercise

T

T&EO	training and evaluation outline
TACSOP	tactical standing operating procedure
TES	Tactical Engagement System
TEWT	tactical exercise without troops
TF	task force
TOW	tube-launched, optically tracked, wire-guided (missile)
TTP	tactics, techniques, and procedures

U

UAC	urban assault course
UO	urban operations

V

VCR	videocassette recorder
VLET	virtual leader effects trainer

W

WARSIM	warfighting simulation
WFF	warfighting function
WFX	warfighting exercise

X

XO	executive officer

Section II. TERMS

bustle rack	storage rack mounted on turrets of some armored vehicles
entity-based model	a representation of an actual or conceptual system that includes mathematical, logistical, or computer simulation
warfighting functions	group of tasks and systems (people, organization, information, and processes) united by a common purpose that commanders use to accomplish missions and training objectives. The warfighting functions are intelligence, movement and maneuver, fire support, protection, sustainment, and command and control. These warfighting functions replace the battlefield operating systems (BOSs).

References

SOURCES USED/DOCUMENTS NEEDED

These are the sources quoted or paraphrased in this publication. All of them are needed by and must be available to the intended user.

DOD Directive

DOD Directive 2311.01E *DOD Law of War Program.* 9 May 2006.

Field Manuals

FM 3-06	*Urban Operations*. 26 October 2006.
FM 3-06.11	*Combined Arms Operations in Urban Terrain*. 28 February 2002.
FM 3-20.12	*Tank Gunnery (Abrams)*. 15 August 2005.
FM 3-21.8	*The Infantry Platoon and Squad*. 28 March 2007.
FM 3-21.9	*SBCT Infantry Rifle Platoon and Squad*. 2 December 2002.
FM 3-21.10	*The Infantry Rifle Company*. 27 July 2006.
FM 3-21.11	*The SBCT Infantry Rifle Company*. 23 January 2003.
FM 3-21.20	*The Infantry Battalion*. 13 December 2006.
FM 3-21.21	*The Stryker Brigade Combat Team Infantry Battalion*. 8 April 2003.
FM 3-21.71	*Mechanized Infantry Platoon and Squad (Bradley)*. 20 August 2002.
FM 3-22.9	*Rifle Marksmanship M16A1, M16A2/3, M16A4, and M4 Carbine.* 24 April 2003.
FM 7-0	*Training the Force*. 22 October 2002.
FM 7-1	*Battle Focused Training*. 15 September 2003.
FM 25-4	*How to Conduct Training Exercises*. 10 September 1984.

Soldier Training Publications

STP 21-1-SMCT	*Soldier's Manual of Common Tasks Skill Level 1.* 14 December 2007.
STP 21-24-SMCT	*Soldier's Manual of Common Tasks (SMCT) Warrior Leader Skill Levels 2-4.* 2 October 2006.

Training Circular

TC 7-98-1	*Stability and Support Operations Training Support Package*. 5 June 1997.

Index

FIELD MANUAL
No. 7-85

HEADQUARTERS
DEPARTMENT OF THE ARMY
Washington, DC, 9 June 1987

FM 7-85
RANGER UNIT OPERATIONS

IMPORTANT

**U.S. Army Infantry School Statement on
U.S. NATIONAL POLICY CONCERNING ANTIPERSONNEL LAND MINES**

TABLE OF CONTENTS

 DISTRIBUTION RESTRICTION: Approved for public release; distribution is unlimited.

PREFACE

This publication contains doctrinal guidelines for the employment and training of ranger units. The intended audience is intermediate and higher level commanders, staffs of those headquarters, and newly assigned officers and NCOs of the ranger regiment. It is to be used with all other published US doctrine concerning NBC operations and the climatic or geographic influences on combat. The fundamentals apply to the employment of ranger units on worldwide operations and across the spectrum of conflict from low to high intensity warfare. This includes all aspects of the integrated battlefield: nuclear, biological, and chemical. It is not intended to apply to the employment of long-range reconnaissance patrol (LRRP) companies, even though such units carry the designation "ranger." These LRRP companies, as well as the long-range surveillance company (LRSC) (corps level) and the long-range surveillance detachment (LRSD) (division level), are organized, trained, and equipped for long-range reconnaissance in enemy territory.

The scope of this publication addresses the mission, organization, equipment, capabilities, limitations, planning, training, operations, and logistical support of ranger units. It describes how the ranger regiment and its subordinate battalions train, receive taskings, plan missions, task-organize, deploy, and integrate supporting elements to perform the mission. Once in combat, the ranger regiment maneuvers to accomplish the mission in the manner of other light infantry units. Ranger units make full use of published light infantry doctrinal guidance. This manual will repeat procedures or techniques contained in other publications only if necessary for understanding or continuity.

This publication does not contain everything the ranger needs to know to successfully execute a unit mission. Rather, it supplements unit training and the ranger unit commander's guidance; it does not supersede it.

The proponent of this publication is HQ, TRADOC. Submit changes for improving this publication on DA Form 2028 (Recommended Changes to Publications and Blank Forms) and forward it to Commandant, US Army Infantry School, ATTN: ATSH-B-ID, Fort Benning, GA 31905.

Unless otherwise stated, whenever the masculine gender is used, both men and women are included.

CHAPTER 1

Introduction

The ranger regiment is a major component of the US Army's special operations forces. It is a unique light infantry unit tasked to conduct special military operations in support of national policies and objectives. These operations require highly trained, well-disciplined units capable of employment in any environment, either alone or in concert with other military forces. Within this publication, the term *ranger force* describes any size force consisting mainly of members of the ranger regiment and led by a member of the ranger regiment's chain of command. A ranger force may be a TOE unit or it may be a specially organized task force for a specific mission.

1-1. Mission.

a. The mission of the ranger regiment is to plan and conduct special military operations. These operations are conducted by specially trained, equipped, and organized forces against strategic or tactical targets in pursuit of national military, political, economic, or psychological objectives. They may support conventional military operations or they may be performed independently when conventional forces cannot be used. (See Figure 1-1.)

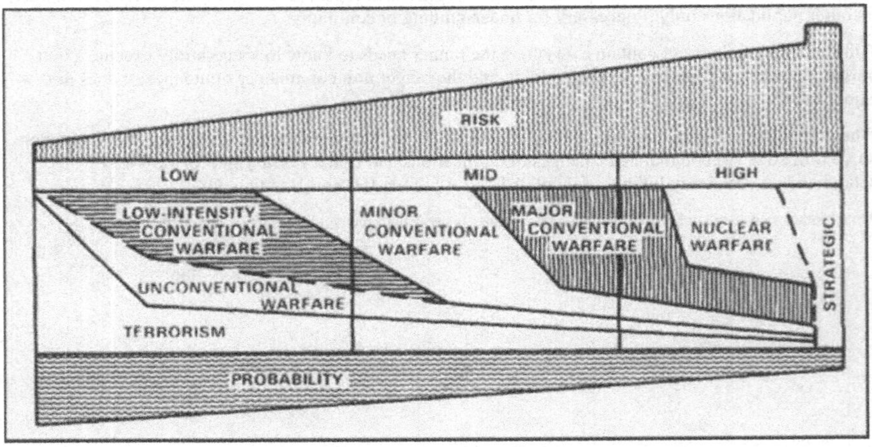

Figure 1-1. Spectrum of conflict.

b. Special military operations conducted by the ranger regiment include strike operations, usually deep penetration, and special light infantry operations. Strike operations include raids, interdiction, and recovery operations. Special light infantry operations include many of the light infantry missions assigned to airborne, air assault, or light infantry battalions and brigades. These operations are conducted in support of the AirLand Battle at all levels of intensity.

1-2. Purpose and function.

The ranger regiment provides the national command authority (NCA) the ability to move a credible military force quickly to any region in the world. The regiment uses the entire spectrum of intelligence support, from national systems to organic assets. Ranger units maintain a readiness posture that supports their immediate commitment to battle once deployed. They are often tailored for specific missions and may require augmentation from external sources. Tactical mobility may be augmented by USAF or Army special operations aviation (SOA) aircraft.

a. The ranger regiment conducts light infantry operations on the integrated battlefield as well as in low intensity conflicts. Special light infantry operations are conducted to accomplish certain contingency missions during peacetime or as part of the echelons above corps (EAC) battle.

b. The regiment conducts both strategic and operational missions. Its efforts are combined into an overall plan to destroy, delay, and disorganize the enemy, or to cause him to divert his attention and combat forces to rear area security. The limited number of ranger units and the diverse targets dictate a careful assignment of missions.

c. The mission, enemy, terrain, troops and time available (METT-T) factors govern the command and control arrangement under which the regiment operates. The ranger regiment or separate ranger battalions are normally assigned to the headquarters whose area of responsibility includes the regiment's operational area(s). The strategic or operational value of potential targets means that ranger units are normally employed at no lower than corps level. Employment at EAC is routine. The EAC headquarters employing ranger units could be the commander in chief's (CINC's) special operations command or field army. The controlling headquarters could also be that of a joint task force (JTF) or the Army component command (ARFOR) of a JTF. The ranger regiment's organization, communications equipment density, and training programs do not support operational level reconnaissance missions.

d. Ranger battalions are not oriented to a specific theater. Current force structure, contingency plans, and training needs (see Appendix G) preclude committing battalions to one region.

1-3. Fundamentals of ranger operations.

a. The success of an operation by a ranger unit depends on the observance of the fundamentals of the US Army's AirLand Battle doctrine (see FM 100-5). The ranger regiment fights the enemy as a light infantry force. It follows infantry doctrine and observes all the basic rules of infantry operations.

b. Along with initiative, depth, agility, and synchronization, ranger operations require--

(1) Detailed planning and coordination that allow the ranger unit to discern and exploit the enemy's weaknesses while avoiding his strength. Both permit the evaluation and use of information gathered from all sources. Detailed coordination integrates all supporting units and services. During planning, the ranger unit conducts pre-mission training, briefings, and rehearsals for all personnel. The use of special mission equipment is practiced and perfected.

(2) Decentralized execution IAW mission orders and the commander's intent. Special operations forces must use individual and unit initiative. Mission-type orders give the ranger force commander the flexibility to take advantage of opportunities on the battlefield.

(3) Surprise, achieved through the ranger unit's ability to move by uncommon means, along unexpected routes, and over rough terrain. Ranger units normally conduct operations during poor weather and reduced visibility, aided by night vision devices. This adds to the attainment of surprise.

(4) Survivability, achieved by using the classic infantry combat techniques of stealth and concealment. The ranger unit engages the enemy at the time and place of its own choosing. It takes full advantage of terrain, and destroys or suppresses enemy weapons. The ranger unit seeks to destroy or neutralize the enemy's command and control systems, his surveillance assets, and his mobility assets. Survivability is enhanced by rapid mission accomplishment and a prompt departure from the objective area.

(5) Mobility, speed, and violence of execution that allow the ranger unit it to close quickly on the objective area and complete the mission before the enemy can react. The speed at which events take place confuses and deceives the enemy as to the intent of the ranger unit. This forces the enemy to react rather than to take the initiative. Tactical mobility lets the ranger unit break contact and withdraw from the objective area.

(6) Shock effect, which is a psychological advantage achieved by the combining of speed and violence with the precision of the ranger attack. The ranger unit strives to apply its full combat power at the decisive time and place, and at the point of the greatest enemy weakness. The ranger unit strives to achieve maximum physical and psychological effect on the enemy by exhibiting aggressiveness and reasoned audacity.

(7) Multiple methods of insertion and attack, trying not to repeat operations thus decreasing the chance the enemy will detect a pattern. This is achieved through imaginative training and planning.

(8) Deception, practiced by the ranger unit during all phases of its operations, from deployment through the insertion phase, to the actions in the objective area and extraction. The enemy is kept unaware of the ranger unit's intentions, is fed conflicting and wrong information, and is kept from knowing the size or mission of the ranger force. The ranger force commander makes full use of ruses, feints, false insertions, electronic countermeasures, and dummy transmissions. Concurrent military operations by US or allied forces may be used as cover for ranger operations. Ranger units are careful to hide every possible aspect of an operation and to disguise those that cannot be hidden. While in the operational area, or when departing any location, ranger units ensure that no material is left that could provide the enemy with information concerning the ranger force.

(9) Audacity, achieved by a willingness to accept risk. The ranger force commander considers what the enemy expects the unit to do and the actual ranger abilities, He then chooses a course of action that may confuse the enemy, while remaining within the capability of the ranger unit. This fundamental is combined with deception and surprise to disrupt the enemy command and control.

c. Detailed planning and coordination finds the enemy. Surprise, mobility and speed, variety, deception, and audacity combine to shock and disorient the enemy, fixing him in place. The violence and precision of the ranger attack finishes the enemy while ensuring the ranger force survives.

1-4. Capabilities and limitations.

a. The ranger regiment has the following capabilities:

(1) Deploying quickly to conduct operations on all types of terrain and in all kinds of weather.

(2) Establishing a credible American presence in any part of the world to show US interest or resolve.

(3) Infiltrating and exfiltrating an area of operations, and assaulting an objective by land, sea, or air.

(4) Conducting strike operations to include raids, personnel and equipment recovery operations, and interdiction of key areas.

(5) Conducting special light infantry operations to include seizing and securing airfields, communications centers, command and control facilities, and key bridges; and other special light infantry operations.

(6) Performing short-duration reconnaissance of assigned ranger objectives for the ranger force commander.

(7) Operating for up to three days without resupply, and for longer periods when provided with accompanying or airdropped supplies.

(8) Providing liaison, communication, and coordination personnel and equipment to integrate the deployed ranger force into the logistical, intelligence, and operational system of the theater or joint task force (JTF) commander.

(9) Assuming operational control, for a limited time, of other US military forces such as engineers or infantry, airborne, or air assault battalions.

(10) Providing the focal point for all-source intelligence support to attached and assigned units of the regimental task force.

(11) Conducting limited combat operations under conditions of chemical, nuclear, or biological contamination.

b. The ranger regiment has the following limitations:

(1) Limited capability against armored or motorized units in open terrain.

(2) No organic transportation.

(3) Limited sustained combat capability due to the shortage of organic combat support and combat service support elements.

(4) Limited organic air defense weapons.

(5) Limited organic indirect fire support.

(6) No casualty evacuation capability.

(7) Reconstitution and retraining needed to replace combat losses.

1-5. Employment considerations.

a. Ranger units are characterized by the quality, motivation, training, and individual skill of their members. This produces units with superb collective abilities, able to adapt well to changing, complex situations.

b. Ranger units can conduct either deliberate or quick-response operations.

(1) Deliberate operations rely on careful planning, reconnaissance and surveillance of the target area, deception, secrecy, thorough preparation and rehearsals, and violent execution. A deliberate operation aims to complete the mission even though the enemy may have heavy forces on or near the objective area. Deliberate operations allow for detailed planning, evaluation, rehearsal, and coordination before insertion. A deliberate operation is likely to succeed against targets that the enemy has protected in depth, that have strong natural defenses, or that need a detailed and long insertion process.

(2) Quick-response operations rely on the high level of training and readiness of the ranger regiment to execute a mission before the enemy can react. These operations are conducted when there is little time for long, detailed planning. They rely on set procedures set forth by the ranger regiment and its supporting elements. A quick-response operation aims to complete the mission before an enemy can react. This type of operation may be chosen due to the time-sensitive nature of the target, political or military goals, the time frame of other operations, or the increased chance of enemy detection.

(3) Whether an operation is to be deliberate or quick response is often a difficult and time-sensitive decision by a high-level command authority. Decision-makers must consider the enemy's strength in the area, his intentions, his ability either to reinforce or to alter the target area, and the consequences of success or failure of diplomatic or military initiatives in related areas. The ranger unit commander tries to use the existing time, manpower, and resources to complete a detailed and coordinated plan. He refines that plan up to the insertion into the objective area.

c. Ranger units train to operate in any environment or weather condition. They regularly perform operations during periods of limited visibility. Ranger units maintain a high state of physical fitness and often train in close quarters combatives. In addition to completing advanced marksmanship training with standard US weapons, each member of a ranger unit trains with many foreign weapons. Ranger units are trained to operate on urbanized terrain, becoming specialists in entry and clearing techniques and quick-fire methods, especially during periods of limited visibility.

d. Because ranger units have limited vehicles, logistics operations capability, indirect fire support, and heavy weapons systems, they are not designed for continuous operations. During all phases of operations and training, ranger units need responsive external support.

e. Ranger units are normally employed against targets and under conditions that need their unique skills. Although targeting priority is set by the overall commander, ranger units are not normally assigned missions that can be done by conventional aerial bombardment or by other units.

f. Ranger units are oriented toward offensive operations. They are not normally employed as a rear area protection force. Although the ranger regiment has a small reconnaissance unit, ranger units do not normally conduct long-range reconnaissance missions. The structure, communications, and training of the ranger unit do not prepare it for LRRP missions. Ranger units engaged in strike and special light infantry operations have a secondary mission to collect and report combat information.

g. Ranger units can be deployed worldwide when US military presence or participation with a host national military activity would serve US interests. This deployment shows a readiness to commit forces into a threatened area or proves US national resolve. After the deployment, other activities include staging operations, rehearsing combat operations, securing base areas for use and deployment of other forces, and so on that provide a clear signal of US intent. Ranger units are not trained or organized to provide mobile training teams (MTT) to train indigenous forces. The US Special Forces or other special operation forces are trained to conduct such security assistance operations.

h. Ranger units can serve as an example to a host country and provide limited military advice and training. They will normally be augmented with linguists and technicians to increase their abilities. The regiment would normally still function as a unit. Most of its assistance would be through short-term, high-impact, unit-oriented operations. They would not be long-term individual efforts associated with advisory-type activities.

i. Ranger units may be deployed to engage in combined training exercises with allies. This enhances US national image by demonstrating the outstanding abilities of the American ranger. These activities may include ranger, light infantry, airborne, air assault, or amphibious operations.

j. Ranger units use standard US nuclear, biological, chemical (NBC) warning; detection; protection; and decontamination equipment and doctrine when operating on the integrated battlefield.

1-6. Command and control.

a. **External.**

(1) The ranger regiment is a key component of the US Army's special operations forces. The other elements of special operations forces are special forces (SF), psychological operations (PSYOP), civil affairs (CA), and special operations aviation.

(2) The 1st Special Operations Command (1st SOCOM) is responsible for command of all active component special operations forces elements, to include the ranger regiment, in peacetime. The 1st SOCOM has command and control of all assigned special operations forces, less operational command (OPCOM) of special operations forces units forward deployed. (See Figure 1-2.) The ranger regiment, while based in the continental United States (CONUS), is under the control of 1st SOCOM. During peacetime, the OPCOM of deployed ranger units is through channels chosen by the NCA through the Joint Chiefs of Staff (JCS) to the unified command.

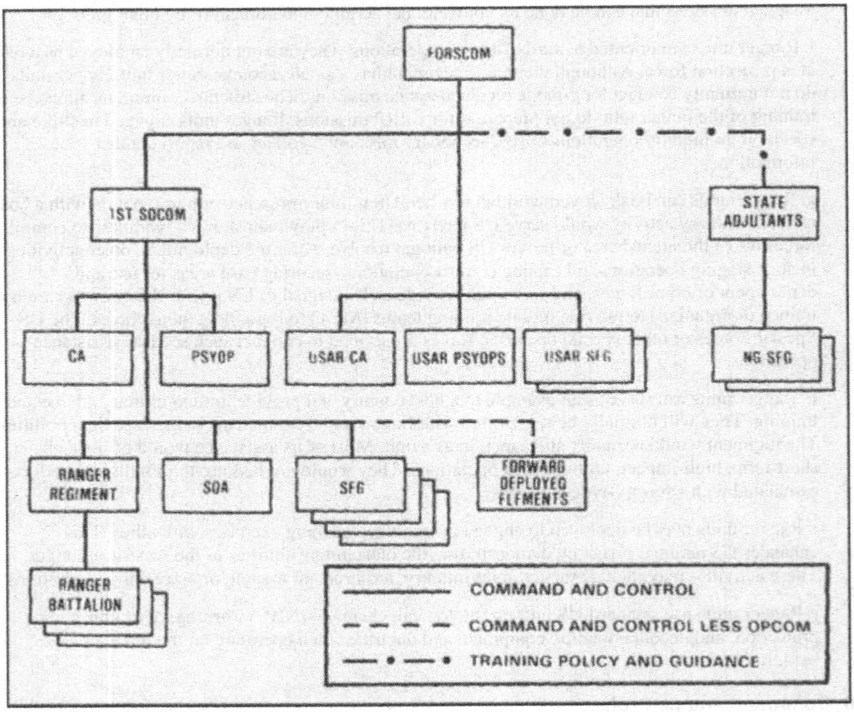

Figure 1-2. Command and control of Army Special operations forces during peacetime.

(3) In support of national policy, the NCA may direct contingency operations involving US

forces. These are politically sensitive military operations normally characterized by the short-term rapid projection or employment of forces in conditions short of conventional war. One aim of the NCA is to conclude them without going to war. The flexibility and effectiveness of the ranger regiment make it possible to use military force quickly to complete the mission and limit the spread of conflict. Command and control of special operations forces in a contingency operation is usually by a special operations task force (SOTF). (See Figure 1-3.) The SOTF and designated special operations forces are under the OPCOM of the highest level of command responsible for the operation. This command may be the unified command responsible for the contingency area (Option A), or a JTF designated by the NCA through the JCS (Option B). If no SOTF is formed, the ranger regiment could be OPCOM to the Army component commander (Option C). The ranger regiment or its battalions are not normally placed OPCOM to a division. A division's area of influence rarely contains appropriate targets. The division also lacks the assets to adequately support ranger operations. Employment of a ranger unit by a corps-level command is normally on a case-by-case basis. The theater or JTF commander retains centralized control over the mission assigned to the ranger unit.

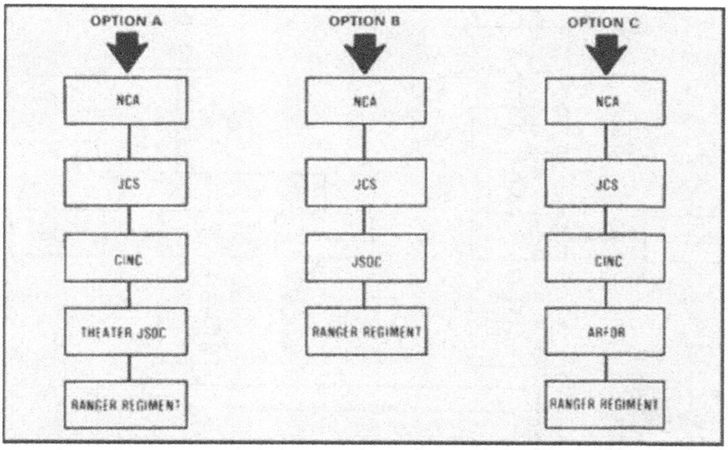

Figure 1-3. Command and control of Army Special operations forces during contingency operations.

(4) The 1st SOCOM is responsible for the deployment of the ranger force to the theater area. The headquarters having OPCOM provides the all-source intelligence, secure communications, insertion and extraction assets, and logistical support needed. The responsible commander in chief, through the Army component commander, provides support during the contingency operation.

(5) During wartime, ranger units are deployed as strategic assets to the theater of operations.

Strategic mission support is provided to the unified commander (CINC) from the NCA through the Joint Chiefs of Staff. It may involve the execution of sensitive operations. The employment of ranger units is controlled at the highest level of command directly responsible for the mission. Ranger units are assigned to a REMAB based on mission, force protection, and OPSEC considerations. The employing command and other elements provide timely and accurate intelligence, communications, logistical, and administrative support. Command and control of these units conforms to the provisions of JCS Pub 2. (See Figure 1-4.)

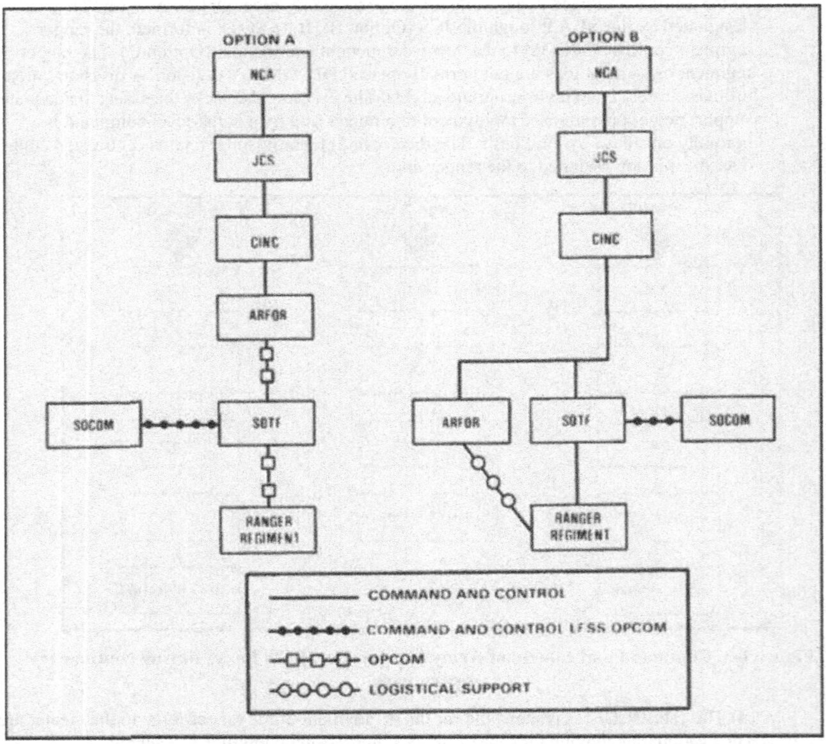

Figure 1-4. Command and control of Army Special operations forces during wartime.

(a) A theater joint special operations command (JSOC) is subordinate to the unified command to supervise the command and control, employment, and support of all US and allied special operations forces. The JSOC is a key element in the command and control of ranger units performing strike missions or special light infantry operations (Option A).

(b) The JSOC may be placed on a lateral line with other in-theater service components (Option B). This can occur when ranger units are -employed far from conventional forces and against strategic targets. It can also occur when the JCS or a CINC forms a JTF to conduct a limited-duration operation in an area where ranger units and conventional forces may operate near each other, or where operations must be coordinated on a regional level.

(c) Under the Army special operations command (ARSOC) concept, 1st SOCOM deploys as a command and control headquarters within the theater. When this occurs, the command relationships will be as shown in Figure 1-5.

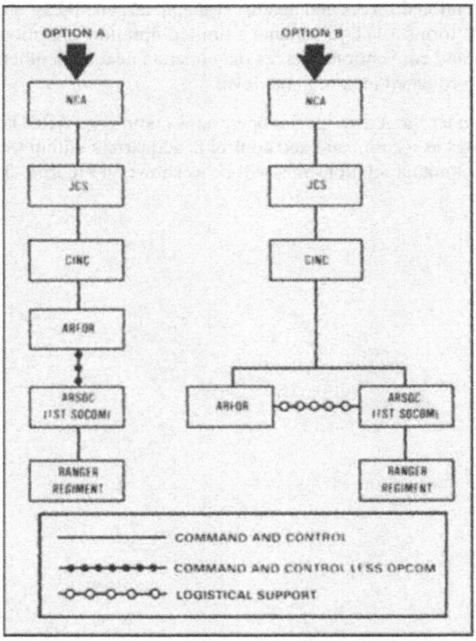

Figure 1-5. Command and control of Army Special operations forces under the ARSOC concept.

Figure –1-5

(d) Recently passed legislation will modify existing high level SOF command and control arrangements. All US Army, USN, and USAF special operations forces will be consolidated into joint commands at both national and theater level. The ranger regiment will continue to operate under the command of the 1st SOCOM, while increasing its ability to conduct joint operations.

(6) Ranger units conducting strike or special light infantry missions are not normally offered for transfer of authority (TOA) to allied forces. However, this will not preclude allied commands and US elements from combined planning and targeting for ranger missions.

(7) Requests for ranger units, from either US or allied commands, go through normal command channels to the CINC. The CINC, through the JSOC, is responsible for the employment of all special operations forces, including ranger units. This provides flexible response to JCS-directed missions, ensures quick response to the CINC's needs, enhances the joint employment of special operations forces, and lessens the chance of overcommitting and degrading the ranger force.

(8) Command and control of the ranger regiment and its battalions is IAW AirLand Battle doctrine. Before commitment to a mission, command and control is kept at a level where the unit's unique skills can be applied worldwide. Once committed, the ranger unit is placed under the operational control of the command that is responsible for the mission. This normally corresponds to the command whose area of influence includes the target area. Ranger task forces may operate directly under the authority of the senior American representative in the country. Command and control arrangements may vary according to the theater involved and METT-T considerations.

(9) The command and control of special operations is centralized and begins at the highest level of the military and civilian decision-making process. The specific command relationships are normally based on the mission.

b. **Internal.**

(1) The ranger regimental headquarters is organized to operate like that of a brigade headquarters. In addition to commanding and controlling all three ranger battalions, the regimental headquarters can assume operational control (OPCON) of conventional combat and combat support units, and other special operations forces for limited periods. It provides the regimental commander with a well-balanced responsive staff. It is structured as an operational headquarters that can deploy to the mission area and act as the ground tactical headquarters.

(a) The regimental headquarters prepares for combat by forming two small command and control groups. Each of these can deploy to an objective area and control combat operations. Other elements of the staff operate the tactical operations center (TOC) and the logistical operations center (LOC). These may be collocated depending on the situation. The TOC and LOC are normally located well behind the FEBA, at the REMAB or at the intermediate staging base (ISB). The ranger regiment also has two small liaison teams consisting of LNOS, staff representatives, and communications elements. (See Appendix B.)

(b) When a single ranger battalion is committed to an operation, the regimental commander normally exercises control from the TOC. A liaison and communications team is detached from the regimental headquarters to augment the ranger battalion. This team is attached to the deployed battalion until the end of the mission.

(c) If two or more ranger battalions are used in a single operation, the regimental commander normally deploys both command and control groups. He commands from the objective area, The use of two small, mobile control elements makes up for the possible loss of the regimental commander and the primary control element.

(d) Whenever the regimental commander is controlling an operation from the REMAB or the objective area, he normally provides a liaison team with a communications element to the higher headquarters exercising operational control. This cell can provide ranger representation to the controlling staff, operations interface, logistics interface, fire support planning and coordination, intelligence analysis and dissemination, and communications between the higher headquarters and the ranger battalion.

(2) The ranger regimental headquarters can serve as a ranger or special operations force task force headquarters. With augmentation, it can function as the Army special operations force command element. The regimental headquarters can also function as the Army component command of a JTF.

1-7. Communications.

a. Ranger missions need unique and dedicated communication support assets. Along with the ranger regiment's organic communications means, communications support assets are organized within the 1st SOCOM that-can deploy in support of the ranger regiment. This ensures that there are reliable communications to all elements of the SOTF. A special operations communications battalion (SOCB), deployed by the 1st SOCOM, provides more communications systems to support the special operations plan. The SOCB does not replace or duplicate existing in-theater communications. It augments the CINC's communications with the US Army's portion of the critical special operations force command, control, and communications link unique to the SOTF.

b. The ranger regiment's communication platoon can provide secure, long-range communication links from the CINC to a deployed ranger force. It can also enter the SOCB and the joint communications support element (JCSE) nets to provide a direct link with the NCA. Its equipment consists of four light vehicles equipped with various multiple communications systems. The two regimental liaison teams use one of these vehicles each. The TOC and LOC each use one.

c. Secure satellite communications (SATCOM) and radios with AM or FM capability are the main means of communication within the ranger regiment. These are also used for communications within the ranger battalions. The regimental commander has two light vehicles equipped with multiple communications systems. One of these is used to support each of the two mobile command and control elements of the regimental headquarters. Some of the radio systems can be dismounted and man-packed.

d. The ranger regimental headquarters has both man-packed and vehicular tactical satellite communication systems. Each of the ranger battalions has one base station and three man-packed systems.

(1) The man-packed radio is battery-operated. It provides two-way communications in both line-of-sight (mobile) and satellite (at-halt) modes. The digital message device group (DMDG) is the input and output device for data transmission. Secure voice communications is achieved using various encryption devices.

(2) The vehicular net control station uses the basic receiver/ transmitter unit installed on a vehicle. Each can serve up to 15 terminals in a communications net.

e. The reconnaissance platoon is equipped with multiple, man-packed communications systems. These let its teams operate in most of the communications nets within the regiment. Each five-man reconnaissance team has this communications ability, as does the platoon headquarters.

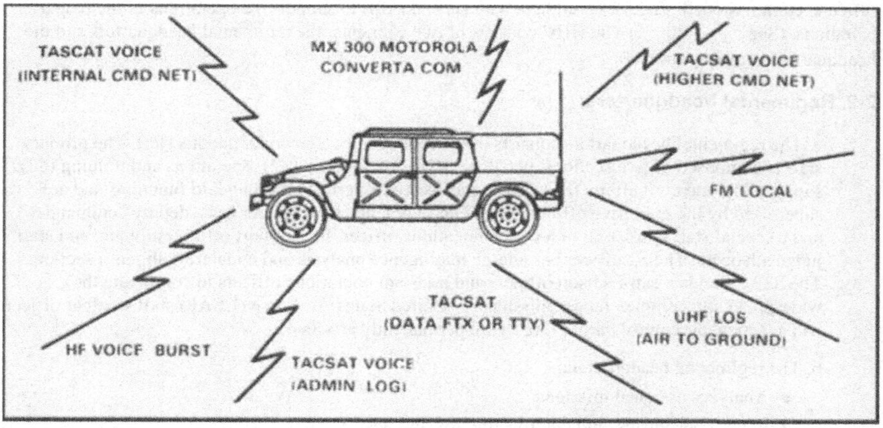

Figure 1-6. Communications capabilities within the regimental headquarters.

CHAPTER 2

Organization and Equipment

The ranger regiment is a specially organized and trained light infantry force consisting of about 2,000 personnel. It includes a regimental headquarters and headquarters company and three ranger battalions. (See Appendix A.) The regiment can operate as a single unit, separate battalions, or separate companies on specific missions. Independent missions are not normally assigned to ranger platoons.

2-1. Headquarters and headquarters company.

The HHC of the ranger regiment consists of 130 personnel, the majority being senior noncommissioned officers, commissioned officers, or soldiers with special skills to support the operations of the ranger battalions. (See Appendix A.) The HHC consists of two elements: the regimental headquarters and the headquarters company (minus).

2-2. Regimental headquarters.

a. The regimental headquarters consists of the regimental commander and his staff. The primary staff is composed of five sections: personnel (S1), intelligence (S2), operations and training (S3), supply (S4), and civil affairs (S5). These staff sections perform the standard functions and are supervised by the executive officer (XO). The regimental commander has a deputy commander and a special staff consisting of a communications officer, fire support officer, surgeon, and staff judge advocate. The S2 officer has a large intelligence analysis and counterintelligence section. The S3 section has extra liaison officers and assistant operations officers to coordinate the widespread and complex ranger missions. A United States Air Force (USAF) staff weather officer and a tactical air control party (TACP) are permanently attached.

b. The regimental headquarters:

- Analyzes assigned missions.
- Plans, coordinates, and controls training and operations.
- Coordinates the infiltration, resupply, exfiltration, and debriefing of the reconnaissance teams.
- Provides liaison support.
- Provides administrative, medical, NBC, and logistical planning and coordination support to the committed ranger battalions.
- Coordinates special support requirements (rigger, transportation, communications security [COMSEC]) for the ranger battalions.

2-3. Regimental headquarters company.

The regimental headquarters company (minus) consists of the company commander and his administrative personnel, plus the limited combat support (CS) and combat service support (CSS) elements assigned to the regiment. These are the fire support element, the communications platoon, the reconnaissance platoon, and the medical treatment team.

a. The company headquarters:

(1) Provides command and control of assigned elements.

(2) Provides administrative, personnel, logistical, and maintenance support to the assigned elements.

b. The fire support element:

(1) Plans fires to support the commander's ground tactical plan.

(2) Coordinates the integration of all available fire support means (mortars, artillery, armed helicopters, close air support, naval gunfire) to support mission execution.

c. The communications platoon (see Appendix A):

(1) Provides secure SATCOM teletype (TTY) and facsimile transmissions (FAX) abilities at the remote marshalling base or intermediate staging base (REMAB or ISB).

(2) Provides secure SATCOM, TTY, and FAX to the regimental headquarters and ranger battalions in the objective area.

(3) Provides cryptographic support to the regimental headquarters.

(4) Provides limited repair of radio and cryptographic equipment.

d. The reconnaissance platoon (see Appendix A):

(1) Trains, equips, and controls the employment of three reconnaissance teams.

(2) Infiltrates the objective area by parachute (high-altitude, low-opening [HALO]; high-altitude, high-opening [HAHO]; static line), helicopter, fixed-wing aircraft, self-contained underwater breathing apparatus (SCUBA), small boat, or other means.

(3) Remains undetected in the objective area for up to five days.

(4) Performs reconnaissance.

(5) Performs demolition target analysis.

(6) Uses small boats and inflatable rafts.

(7) Emplaces and monitors unattended ground sensors.

(8) Uses night image enhancement devices.

(9) Uses photographic equipment.

(10) Emplaces electronic target acquisition and designation devices.

(11) Collects specific combat information to satisfy priority intelligence requirements (PIR).

(12) Reports information using secure, long-range communication equipment.

Figure 2-1. Reconnaissance platoon communications.

(13) Performs drop zone (DZ) and landing zone (LZ) selection, marking, and reception duties.

(14) Escapes and evades enemy areas.

(15) Performs limited attacks or ambushes when tasked by the regimental commander.

(16) Reports weather in the objective area.

e. The medical treatment team:

(1) Provides medical support to the regimental headquarters.

(2) Coordinates required medical support and evacuation.

2-4. Ranger battalion.

a. The ranger battalion is organized and equipped to conduct both special and conventional combat operations. It contains organic combat, CS, and limited CSS elements. It consists of a battalion headquarters, a headquarters company, and three ranger rifle companies. (See Appendix A.)

b. The ranger battalion is the combat element of the ranger regiment. It can conduct independent combat operations or subdivide into its separate ranger rifle companies and their subordinate elements who can conduct semi-independent operations.

c. The ranger battalion headquarters and headquarters company (see Appendix A) includes:

(1) **Battalion headquarters.** The battalion headquarters consists of the commander; the executive officer; the command sergeant major; the S1, S2, S3, S4, and S5 officers; the chaplain; the communications officer; the fire support officer; and two radio operators.

(2) **Battalion headquarters section.** The battalion headquarters section consists of the S3 air and assistant S8, an assistant S2, an intelligence sergeant and a senior intelligence analyst, two operations sergeants and an operations assistant, a personnel staff noncommissioned officer (PSNCO), the chaplain's assistant, two clerk typists, two radio operators, a chemical officer and NCO, a personnel administration center (PAC) NCO, three PAC clerks, and a legal specialist. There is a USAF tactical air control party permanently attached.

(3) **Communications section.** The communications section consists of one telecommunications technician, one tactical communications chief, two combat signal radio team chiefs, one tactical communications systems supervisor, three radio operators, one switchboard operator, one single channel radio operator, and two equipment mechanics. The section provides communications support and associated maintenance.

(4) **Battalion medical section.** The battalion medical section consists of one general medical officer, one physician's assistant, two medical specialists, and twelve aidmen. The section provides unit-level medical support and trains selected members of the battalion in advanced medical procedures.

(5) **Battalion support platoon.** The battalion support platoon consists of a battalion support platoon leader, an ammunition NCO, and a food service section. The platoon provides limited CSS for assigned and attached units. This platoon does not have organic transportation. The battalion food service section consists of a food service sergeant (E7), a first cook (E6), and eight cooks.

(6) **Fire support team headquarters.** The fire support team (FIST) headquarters is supervised by the battalion fire support officer. It consists of the fire support sergeant and two fire support specialists. The headquarters controls three 9-man fire support teams. These teams are normally with the ranger rifle companies in a habitual support relationship. When operations are conducted near water, the fire support team headquarters may be augmented by spotter teams from the naval shore fire control party.

(7) **Company headquarters.** The HHC company headquarters consists of a commanding officer, an executive officer, a first sergeant, a supply sergeant, and an armorer.

d. The ranger rifle company (see Appendix A) includes a company headquarters, three rifle platoons, and a weapons platoon. Each company can conduct semi-independent operations for a short time when augmented by assets from the ranger battalion headquarters company.

(1) **Ranger rifle company headquarters.** The ranger rifle company headquarters consists of a commanding officer, an executive officer, a first sergeant, a supply sergeant, a tactical communications chief, two radiotelephone operators, an armorer, and an NBC NCO. It also has a two-man sniper team equipped with a sniper weapon. It normally has one of the 9-man FISTs attached from the ranger battalion headquarters company. Elements of this FIST may be further attached to the ranger rifle platoons.

(2) **Ranger rifle platoon.** The ranger rifle platoon consists of a headquarters section, three rifle squads, and a machine gun squad. The headquarters section personnel include a platoon leader, a platoon sergeant, and a radiotelephone operator. Each rifle squad consists of a squad leader and two balanced fire teams of four men each. The machine gun squad consists of a squad leader and three M60 machine gun crews of three men each. The machine gun squad is normally employed as a single element to provide quick, accurate, long-range supporting fire. One or more crews may be attached to other squads within the platoon.

(3) **Ranger weapons platoon.** The ranger weapons platoon consists of a headquarters section, a mortar section, and an antiarmor section. The headquarters section consists of a platoon leader, a platoon sergeant, and a radiotelephone operator. The mortar section has two 3-man 60-mm mortar squads. The mortar squads are normally employed as a section. They may be attached by section or squad to other platoons. The antitank section normally consists of a section sergeant and three 3-man 90-mm recoilless rifle or Dragon missile teams. They are normally employed as a section. The antiarmor section may also be attached by section or team to other platoons.

2-5. Special teams and elements.

The ranger regiment forms special teams to do critical tasks. These teams are made up of members of the unit and are used when needed. Examples of these special teams and elements are--

a. SCUBA teams whose members are graduates of the Special Forces Underwater Operations School. They conduct special insertion and reconnaissance duties.

b. HALO teams whose members are graduates of the Special Forces Freefall School and who conduct high-altitude, low-opening parachute operations.

c. HAHO teams whose members are also graduates of the Special Forces Freefall School and who conduct high-altitude, high-opening parachute operations, allowing offset parachute insertions.

d. Scout swimmer teams whose members are specially trained by the Landing Force Training Command (United States Marine Corps [USMC]) to conduct operations along coastlines, waterways, and inland waters. These teams conduct special insertion, reconnaissance, and security tasks often in support of amphibious operations.

e. Sniper teams, whose members are authorized by the table of organization and equipment (TOE), are used for both conventional and special operations. Along with the two-man sniper teams in the rifle company headquarters, each rifle platoon has two trained snipers.

f Demolition teams whose members are trained in advanced demolition and incendiary techniques and the use of special explosives,

g. Air defense teams trained in the use of man-portable air defense system (MANPADS) weapons. Such teams give the ranger battalion limited self-defense against air attack.

h. Liaison teams that are used by the ranger regimental commander to augment the existing command and control links from the employing unit headquarters to the ranger regiment. These teams are placed where they are most needed and may assist the higher headquarters staff in planning and target selection, as well as performing other liaison duties. See Appendix B for a detailed listing of the personnel and equipment included in the ranger liaison teams.

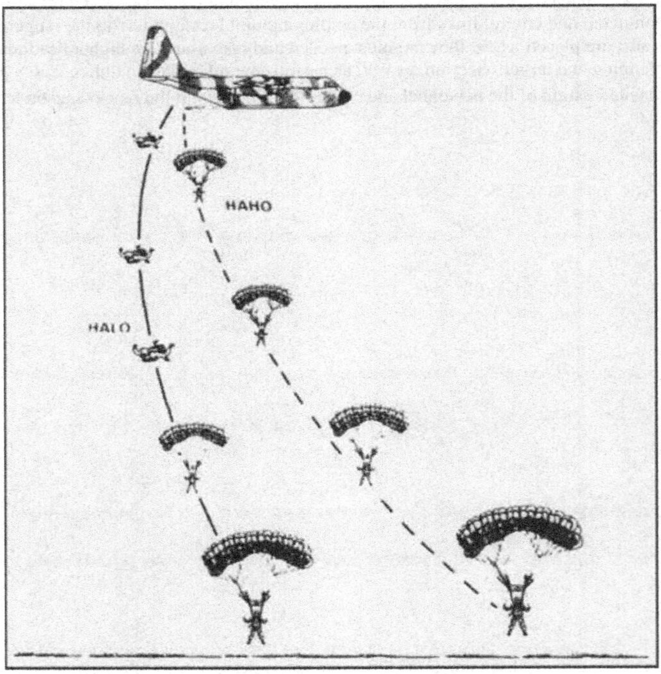

Figure 2-2. HALO and HALO parachute operations.

Figure 2-2

i. The ranger support element (RSE).

(1) The RSE is a special support element provided by host installation commanders at CONUS posts. It consists of elements and teams from either TOE units stationed at the installation or from the table of distribution and allowances (TDA) organizations under the command of the installation commander.

(2) The RSE provides the necessary support to ensure a timely and efficient deployment of the ranger battalion. During CONUS deployments, the RSE is responsible for both the departure and the reception support provided the battalion. During OCONUS deployments, the RSE is responsible for initial deployment and continuing support until the deploying ranger force is OPCON to the supported unified command or JTF. The responsible theater commander assumes support of the ranger battalion at the REMAB or the ISB if outside continental United States (OCONUS). On deployments that require the ranger battalion to move directly to the objective area from CONUS, the RSE continues support of the ranger battalion until released, and is prepared to accept considerable augmentation assets. The RSE provides the following support:

(a) Furnishing meals that are either served by a supporting unit in an established facility or taken to the ranger battalion at an isolated location.

(b) Issuing supplies from existing stocks, or from contingency items stored for use by the ranger battalion.

(c) Transporting supplies, food, personnel, and equipment.

(d) Palletizing equipment and ammunition for air movement and airdrop.

(e) Rigging vehicles and equipment for airdrop.

(f) Refueling vehicles and equipment.

(g) Issuing barrier and construction material for building of rehearsal sites.

(h) Helping in the construction of rehearsal sites and targets.

(i) Securing, storing, moving, and issuing ammunition and other Class V items.

(j) Maintaining ranger weapons and equipment, to include communications equipment. This is normally done by maintenance contact teams.

(k) Setting up and operating the RSE emergency operations center.

(l) Securing the REMAB if it is in CONUS.

CHAPTER 3

Deployment

Ranger units can quickly deploy by any means of transportation to any location in the world.

3-1. Readiness.

One ranger battalion is always in an advanced readiness condition as the Ranger Ready Force (RRF) available for immediate worldwide deployment. A second battalion is prepared to deploy later and a third battalion to follow. The regimental headquarters maintains command and control, liaison, communications, and reconnaissance elements immediately available for deployment. Higher status of readiness in response to specific world situations can be achieved.

3-2. Ranger ready force.

In order to maintain advanced readiness conditions, ranger battalions are rotated as the RRF.

a. Before taking over the duties of the RRF, battalions must do the following:

(1) Complete preparation for oversea movement (POM) qualification, to include an individual records check and required inoculations.

(2) Test fire all weapons, confirm battlesight settings, and repair or replace faulty weapons.

(3) Assemble and load on pallets the unit basic load (Classes I and V) and other supplies and equipment.

(4) Check unit-derived packing list components for accountability and serviceability.

(5) Prepare special operations troop listings.

b. During an RRF period, ranger units:

(1) Have a positive recall system for all personnel.

(2) Store enough air items for a battalion-size parachute assault.

(3) Conduct no off-post exercises, unless approved by the regimental commander.

c. Each unit keeps an alert roster and a telephone notification system for recalling personnel. The rosters ensure that the duty officer and the unit charge of quarters make a minimum number of calls. Each commander briefs personnel assigned to his unit on the recall system and tests it often.

d. Ranger units can maintain an increased readiness condition for long periods, but they should not be kept in that condition longer than needed. The effort to maintain this level of readiness will degrade the unit's training and effectiveness. If the needs of the operations dictate long periods at the highest readiness condition, provisions must be made for sustainment training. Also, added support will be required from the RSE.

e. The ranger regiment maintains a strong family support system when the unit is training or deployed on combat operations. Each member of the chain of command is responsible for family support.

3-3. N-hour sequence.

The N-hour sequence is a departure time sequence designed to help the ranger force complete all required deployment actions within a certain time. The length of the N-hour sequence depends on the readiness condition of the battalion before deployment. The N-hour sequence begins when the battalion is alerted and ends when the first elements depart. The N-hour sequence does not include travel time to the objective area.

3-4. Intelligence and security.

a. **Essential elements of friendly information.** Items of sensitive information and subject areas, if not secured, can be collectively analyzed by a hostile force. This analysis can reveal the intentions of a unit or compromise a mission. These items of information are designated as essential elements of friendly information (EEFI). They are never transmitted without encryption. The EEFI associated with a ranger unit alert normally are the mission, the use of staging bases, nature of the target, special terrain or weather conditions, and time limits. The ranger force commander must ensure that EEFI are not discussed outside of secure areas or made available to personnel without a security clearance and a need to know.

b. **Security plans.** Security plans for billets, marshalling areas, and out-loading areas must remain flexible to adapt to any situation. Certain items are always addressed in security plans. These are access control and physical security.

(1) **Access control.** Ranger units control access to their respective areas by allowing only authorized personnel to enter or leave. All entrances are secured and an armed guard with the access roster is posted. Access to telephones is restricted.

(2) **Physical security.**

(a) Equipment is open to theft and sabotage during the early phases of deployment. Units must provide guards to secure all unit and personal equipment.

(b) Weapons, ammunition, and sensitive items must be stored in a restricted area and guarded.

(c) Classified documents are open to compromise during deployment. Steps must be set up for maintenance, transportation, turn-in, and destruction of classified material. The most secure facility available must be used.

c. **Rear detachment security.** Security after deployment (maintained by the rear detachment) is as important as security before deployment. No information should be given to the news media from other than official sources. The rear detachment may continue to play a role in the unit cover and deception plan.

d. **Tactical operations center security.** Only personnel listed on TOC access rosters are allowed free entry into the TOC area. Personnel wanting to enter the TOC, who are not listed on the access rosters, are detained until their identity and purpose are proven. Such personnel are escorted at all times while within the TOC area.

e. **Marshalling and out-loading area security.** The ranger unit normally marshalls its personnel and equipment in the unit area, providing its own security force. The USAF security police provide this support at the departure airfield. On Army airfields, Army military police provide security. En route security is provided by the RSE.

f. **Personnel and unit equipment sanitization.**

(1) Before deployment, personnel are relieved of non-mission-essential material having intelligence value. This reduces the intelligence available to hostile forces if they capture a ranger force member.

(2) The ranger force commander sets the degree of sanitization IAW mission requirements.

(3) If strict sanitization is imposed, guidelines are issued as follows:

(a) The only personal items retained are identification card and tags.

(b) Unit insignia and markings are removed from uniforms and equipment.

(c) Personal items, such as wallets and photographs, are secured by the rear detachment commander.

(d) Personnel are briefed on how to compose and retain combat diaries. No personal diaries are allowed.

(4) The commander decides how much US currency is to be taken with the unit. '

g. **Area sanitization.** To avoid the compromise of classified or sensitive mission-related material, searches are conducted before and after leaving the host installation, REMAB, or ISB. These searches are held to locate and remove any material, indicators, litter, or other items that could compromise the ranger unit's mission. Unit commanders ensure that a thorough sanitization of their areas is conducted, to include dismantling rehearsal areas, destroying terrain models, erasing chalkboards, and emptying all classified and nonclassified waste containers. During their searches, commanders must pay close attention to emergency operations centers, tactical operations centers, and marshalling and loading areas. The ranger unit's intelligence officer coordinates support from local counterintelligence elements to search the area again after the ranger unit leaves.

3-5. Operations.

a. **Actions upon receipt of the OPORD.**

(1) Task-organize and equip units for special mission(s) IAW the warning order.

(2) Allocate/reallocate mission-essential equipment.

(3) Distribute augmentee personnel.

(4) Redistribute key personnel,

(5) Prepare a deployment time sequence.

(6) Prepare a movement plan.

(7) Prepare and maintain a current operation estimate.

b. **Advance elements.** A planning element is normally deployed to the REMAB/ISB or JTF headquarters in advance of the main body if the ranger force does not deploy directly to the objective area.

c. **Emergency operations center.** An EOC is setup by the ranger battalion headquarters. The EOC is responsible for relaying all reports to higher headquarters. It remains operational until ordered to cease.

d. **Air movement.** The ranger force always deploys by USAF aircraft. This is normally an administrative air movement to a REMAB or ISB. It may be followed by tactical air deployment from there. For some missions, tactical deployment begins at the unit's home station. Even if the movement is administrative, key personnel are cross-loaded. This is done under the direction of the operations section air officer based on aircraft allocations and load plans.

3-6. Logistics.

The following describes logistics readiness needs and steps for deployment.

a. **Supply.**

(1) **Class I.**

(a) Combat rations are normally palletized when the unit is RRF. These rations are to be used during the tactical operation. If the mission calls for deployment directly into the objective area, the palletized stock of rations is off-loaded in the unit area and issued as needed.

(b) Each ranger deploys with emergency combat rations.

(2) **Classes II and IV.**

(a) **Individual clothing and equipment.**

- Individual A bags are packed IAW a predetermined packing list, tagged, and marked. They are palletized, moved to the departure airfield by the RSE, and loaded IAW the air movement plan.
- Individual B bags are packed, assembled, and forwarded on order by the rear detachment.

- Individual environmental equipment is issued. Equipment shortages are replenished out of stocks maintained by the RSE.

(b) **Air items.**

- While the unit is on RRF, a contingency load of main and reserve parachutes is maintained by the RSE.
- Individual and unit air items are palletized by the RSE and shipped in bulk to the REMAB or ISB.
- For direct-deployment missions, air items and packing material are taken to the unit by the RSE.

(c) **General supplies.** The RSE maintains pallets of common contingency items and special equipment. Pallets are picked up by the out-loading support unit and moved to the unit area for issue or moved with the other pallets to the departure airfield.

(d) **Bulk supplies.** Barrier material is identified, relocated, palletized, and out-loaded IAW the air movement plan.

(3) **Class V.**

(a) The unit basic load of ammunition is secured, palletized, and loaded by the RSE IAW the air movement plan.

(b) For direct deployment missions, selected pallets are taken to the unit area and ammunition is issued. The convoy route is coordinated through the post EOC, and secured by the RSE.

(c) Designated items are rigged by the RSE for airdrop, if enough time exists before movement to the REMAB or ISB begins.

(4) **Class VI.** Personnel deploy with comfort items as needed.

(5) **Class VII.**

(a) All weapons and sensitive items not deployed are assembled and secured by the rear detachment personnel.

(b) Unit vehicles are prepared for airland or airdrop.

(c) Inoperable mission-essential equipment is repaired or replaced.

(d) Special equipment requests are forwarded to the logistics section as they are identified.

(6) **Class VIII.**

(a) Individuals deploy with one set of individual Class VIII items.

(b) Unit medics deploy with modified aidbags.

(c) Controlled medical items (less drugs) are palletized and deployed with section supplies. Controlled drugs are secured and issued to unit aidmen on order.

(d) Medical supply chests are maintained in the unit aid station.

(7) **Class IX.**

(a) Units deploy with initial prescribed load list (PLL) items.

(b) Critical PLL shortages are identified to the RSE for follow-on issue.

b. **Maintenance.**

(1) Upon alert, priority maintenance or replacement of key items of equipment is coordinated by the logistics officer. Only minimum mission-essential equipment is replaced.

(2) Items to be returned from supporting maintenance activities are identified, and pickup is coordinated by the property book officer.

c. **Out-loading coordination and support.**

(1) Upon alert, the RSE provides a liaison officer to the ranger logistics section.

(2) The support platoon leader (SPI,) attends the air movement conference and coordinates the following:

(a) Movement times (for both troops and equipment) to the departure airfield.

(b) Steps for weighing and joint inspection of the pallets and vehicles upon arrival at the departure airfield control group (DACG).

(c) Limitations on the pallet height and weight due to aircraft type, allowable cargo load (ACL), and availability.

(d) Changes in the planned number of pallets due to the mission requirements or airframe availability.

(e) Rigging and loading of the heavy drop and supply bundles.

(3) Empty pallets are placed in position for loading by the RSE in coordination with the SPL.

(a) Companies and staff sections prepare pallets.

(b) For airdrop operations, vehicles are prerigged at the unit location. On order, vehicles are driven to the heavy drop rigging site. The SPL and the officer in charge of the rigger detachment coordinate for a rigging site and the delivery of the required air items to that location. Rigging is done by the RSE, and supervised by a member of the ranger unit. Loading aboard the aircraft is IAW the air movement plan and supervised by RSE and USAF aerial port personnel.

(4) The DACG is provided by the host installation. Upon arrival at the DACG area, unit personnel and equipment are under the control of the DACG. As the deployment begins, a

member of the ranger unit is present at the DACG to ensure an orderly loading of personnel and equipment.

(5) The DACG plans for troop-holding areas in the event of delayed or aborted aircraft.

(6) Unit movement data are maintained by the air officer in the operations section.

3-7. Communications-electronics.

a. **Alert notification communications.** Alert notification of the ranger battalions by the regiment is done by secure communications means. Units in local training areas are informed by radio or through their assigned range field telephone. Units operating out of normal FM radio range during RRF periods set up a high frequency net, or other communications, to their higher headquarters.

b. **Telephone service.** Strict control is maintained over key telephones being used during unit alert. Care is taken to ensure that COMSEC and OPSEC violations do not occur. All telephones are assumed to be monitored; classified, sensitive, or official information is not discussed.

c. **Radio communications.**

(1) **Voice communications.**

(a) A secure radio net is set up early in the deployment. The ranger unit EOC, the installation EOC, and the RSE EOC operate in this net. A liaison officer from the ranger unit is placed at the installation EOC.

(b) The ranger support element liaison officer, attached to the logistics section, operates a secure voice station on a separate logistics net to the RSE EOC and the RSE commander.

(2) **Cryptographic material.** During RRF, COMSEC codes and enabling equipment are kept available to set up the secure voice net.

3-8. Remote marshalling base and intermediate staging base.

a. Ranger units can deploy from their CONUS base directly to the objective area. A more common method would be for the ranger unit to deploy first to either a REMAB or an ISB before insertion into the area of operations. (See Figure 3-1.)

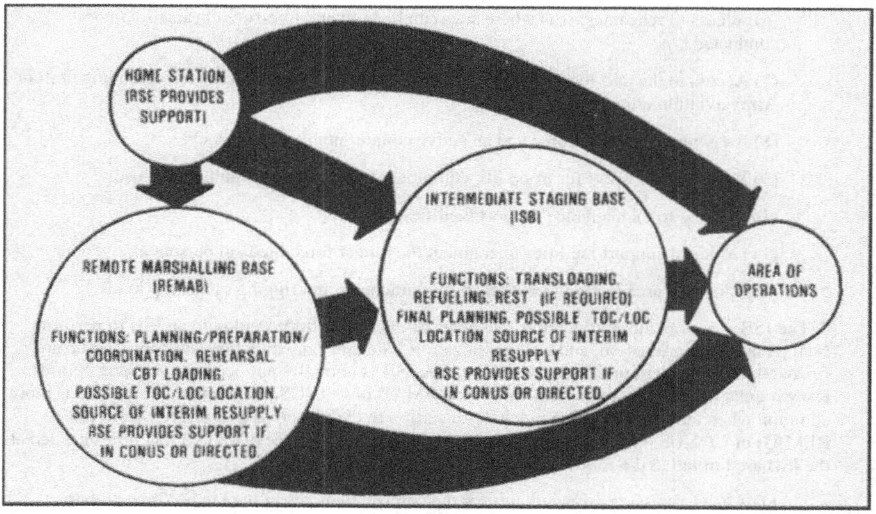

Figure 3-1. Deploying ranger units.

b. The REMAB is a secure base to which the entire ranger unit, including organic and attached support elements, deploys and where mission planning continues. The REMAB is normally within the geographical area encompassed by the command authority of the theater or JTF commander. This ensures that the CSS elements providing support to the ranger unit are operating within their normal area. It prevents or lessens out-of-sector support requirements for CSS elements.

c. The REMAB also provides a secure location for the ranger unit to conduct detailed planning and coordination with the controlling headquarters staff. In the REMAB, the ranger force commander conducts rehearsals, refines and modifies plans, determines PIR, and coordinates with the proper intelligence source to receive that information. It is in the REMAB that specialist augmentees to the ranger force are integrated into the unit, if they have not already joined. Specially trained supporting units, such as aviation and communications elements, also join the ranger force at the REMAB. The REMAB must provide--

(1) Access to the controlling headquarters staff.

(2) Physical security of billeting, planning, maintenance, and communications areas.

(3) Mess, billeting, latrine, and shower facilities for the ranger force and its supporting elements.

(4) Access to a C-141 or C-130 capable airfield, possibly with all-weather operations.

(5) Access to secure communications and processed mission intelligence.

(6) Access to rehearsal areas where sites can be built and live-fire rehearsals can be conducted.

(7) Access to the unit locations of major supporting elements, such as naval landing craft or Army aviation units.

(8) An external security force and an active counterintelligence agency.

(9) Vehicular transport for troop lift, equipment transfer, and administrative use.

(10) Access to maintenance support facilities.

(11) Medical support facilities to augment the ranger force medical personnel.

(12) Covered area to allow for packing of parachutes and rigging of airdrop loads.

d. The ISB is a more austere location. Elements of the ranger force deploy to an ISB to perform final planning, coordination, and task organization. The unit's organization and composition are finalized for movement to the objective area. The ISB is normally not occupied for long periods. Ranger units may deploy to the ISB from the REMAB or CONUS. Deployment to the ISB is more common when terrain or distance precludes insertion to the objective area directly from the REMAB or CONUS. Quick-reaction missions often involve the use of an ISB. Facilities needed at the ISB are limited to the following:

(1) Austere messing arrangements for the combat elements of the ranger force and its attachments.

(2) Austere billets or rest areas.

(3) Water point.

(4) Fuel for aircraft and vehicles.

(5) Areas for test firing of weapons.

(6) Covered and concealed areas for assembly of the ranger force and rigging of parachutes and door bundles.

(7) Austere airfield support facilities, possibly capable of all-weather operations.

(8) Remote location far away from civilians or traffic routes.

(9) Security and counterintelligence elements.

(10) Secure communications.

e. The REMAB and ISB should be located in an area similar in terrain and climate to the objective area. Time spent at the REMAB or ISB lets the ranger force begin acclimatization. Ranger units train yearly in all climates and emphasize physical conditioning. This allows acclimatization quicker than units not as well trained or conditioned. However, ranger units still need an acclimatization period if the terrain and climate of the objective area are different from that of their CONUS base. Not enough acclimatization means that the ranger force could be less efficient, especially when the objective area is hot, dry, or at a great altitude.

CHAPTER 4

Insertion, Extraction, Escape, and Evasion

Insertion is the entrance of a ranger force into a hostile area to execute a particular mission. It is the first critical phase of a ranger operation. These operations often involve deep penetrations of enemy territory by air, water, or land. This chapter emphasizes fundamentals and techniques employed by ranger units when their mission requires insertion and extraction. Extraction is conducted quickly after the mission to avoid casualties.

4-1. Intelligence.

a. Insertion plans are based on timely and accurate intelligence. The headquarters directing the operation must provide current intelligence information to the ranger force. The ranger regiment normally takes action to solicit required information from the controlling headquarters and from other agencies, if security permits. The type of intelligence needed varies depending on the specific mission, but normally includes the following:

(1) Current photographs from aerial or overhead reconnaissance.

(2) Detailed maps.

(3) Data on the population, terrain, and weather from area studies.

(4) Current enemy order of battle, enemy troop locations, and enemy capabilities.

(5) Details on the target area and infiltration routes.

b. To meet its need for current intelligence, the ranger force also relies on aerial reconnaissance and surveillance of the terrain over which it will move. Emphasis is on getting information on the enemy's ability to detect and engage the inserting forces. The location and abilities of air defense radar and weapons systems are critical. Terrain analysis is detailed and focuses on areas suitable for the insertion. Weather information must be detailed and current.

c. In getting information and intelligence data, OPSEC measures are stringent and must be enforced. Coordination with other agencies is kept within security constraints approved by the directing headquarters.

4-2. Deception.

There are always plans to deny the enemy knowledge of the ranger unit's insertion, or to deceive him as to the location or intent of the operation. Feints, false insertions, and other deception operations all add to ranger deception plans. Selection of unexpected means, time, place, and routes of insertion, as well as use of speed and mobility, give less reaction time to the enemy. Planning may include the use of diversionary fires to direct the enemy's attention away from the inserting unit. Deception techniques that may be used include--

a. Multiple airdrops, water landings, or both. This precludes the entire force from becoming engaged at the same time and place, if detected.

b. Dispersion of insertion craft (air or water) both in time and location.

c, Landing a force in an area closer to other potential targets than to the actual targets.

d. Leaks of false information and false messages.

e. False insertions or radio traffic to disguise actual insertions.

f. Diversionary actions such as airstrikes in other than the target areas.

g. Increased reconnaissance flights, either by manned aircraft or remotely piloted vehicles, over false targets.

4-3. Speed and Mobility.

a. Speed of action must be emphasized when the enemy can react quickly to the ranger force presence or when there are limited means of withdrawal. The operation must be conducted swiftly, allowing the ranger force to withdraw before the enemy can react.

b. Major problems facing ranger units are the need for speed and mobility, and the heavy loads carried by soldiers for sustained operations. Ranger units are required to be flexible and to be able to react to any problem within the objective area. However, if equipment is carried to cover every encounter, the load will become too heavy. This will reduce the speed and mobility of the ranger unit. The ranger force commander must always have a balance between the equipment carried and the mobility of his unit. He does this by making three different types of loads: fighting, mission, and existence. He decides what items will make up these loads, where they are carried, and what will be done with them upon contact with the enemy.

(1) The fighting load consists of those items of equipment, weapons, ammunition, food, and water that are common to all soldiers and are necessary for immediate use in combat. These items are normally carried and are retained at all times while in contact with the enemy.

(2) The mission load consists of those items of equipment or ammunition that are required by the mission, but are not normally carried by all rangers in a unit. The load may include spare batteries, demolitions, radios, antitank weapons, or mines. This load is normally spread out among members of the unit and is rotated often. It may be dropped upon enemy contact, but is normally carried to the objective and either used or emplaced there.

(3) The existence load consists of any items designed to provide protection from the elements: sleeping gear, changes of clothes, spare rations, or tentage. This load may or may not be brought into the objective area, depending on the tactical situation. If it is brought, it is normally dropped upon enemy contact or left in an assembly area and retrieved later.

4-4. Stealth.

Methods and steps chosen for insertion must emphasize stealth and try to avoid detection by the enemy. Multiple routes or methods of insertion may be used to preserve the ranger force if some soldiers or units are seen.

4-5. Suppression.

Effort must be made to suppress enemy detection devices, weapons systems, and command and control facilities. This may include electronic countermeasures or direct attack. This helps prevent the enemy from finding and engaging the ranger force en route to the objective area. Deception tactics add to suppression activities. If the ranger force is inserting by infiltration, suppression must not disclose friendly activities.

4-6. Security.

Emphasis on OPSEC prevents compromise of an impending operation. The operation must not be disclosed by overt rehearsals or training activities, the open use or procurement of special items of equipment, or the location of the marshalling area. Several measures can help in maintaining security:

a. Guarded and secure facilities for the headquarters and staff during planning.

b. Issue of operation orders only after units are moved to secure marshalling areas.

c. Alerts, rehearsals, and training to get the local people used to conditions that will exist during actual marshalling.

d. A secure facility set up to isolate units selected for an impending operation.

4-7. Night vision and electronic devices.

a. Rangers use night vision devices to detect and avoid the enemy's forces and his detection devices. These devices also assist in controlling and speeding movement and allow units to traverse almost impassable terrain.

b. The use of electronic homing devices permits precise navigation during reduced visibility. These devices may also be used with visual marking systems.

c. Passive night vision devices are used to assemble rapidly and to reorganize.

d. The use of active night vision devices in assembling, and the degree to which drop zones or assembly areas are "sterilized," depends on the following:

(1) Mission success may depend on not being seen during insertion. When a clandestine insertion is made, the time needed to get to the objective will normally be greater, the use of passive devices will prevail, and the sterilization of the area will be vital. Each ranger must be briefed on sterilization plans and techniques for erasing signs of the insertion.

(2) When detection is likely and the mission depends on getting to the objective before the

enemy can react, speed becomes crucial to the success of the insertion. Insertion may be near or on the objective. Rangers must get in and out quickly. They do minimum sterilization of the area and can use active RSTA devices.

4-8. Rehearsals.

Rehearsals should be close to the actual conditions of insertions and extractions. Security of rehearsals and rehearsal sites is important to the success of the operation. Rehearsals are held on terrain and structures like those in the objective area. If such structures are not available, they are built by the ranger force and the RSE with its supporting engineers.

4-9. Sand tables and terrain models.

Sand tables are useful to acquaint personnel with insertion sites and surrounding terrain. The use of sand tables and terrain models during the issuance of prejump orders and briefings enhances an orderly and speedy tactical assembly.

4-10. Communications.

Reporting during insertion and extraction missions is normally by exception, using an operations schedule.

4-11. Planning insertions.

a. **Centralized planning.** Planning for insertion of ranger units is centralized at the lowest level responsible for the overall operation. This is normally at the joint task force level, with detailed coordination required among all supporting forces. These forces may include other Army elements, USAF, United States Navy (USN), United States Marine Corps (USMC), other US agencies, and allies. Preparation must include detailed contingency planning for use of firepower, shock effect, speed to break contact (if detected), control measures, procedures, and planning. When planning takes a long time, intelligence must be continually updated through all available sources. This includes overhead photography and ground sources to ensure that the initial planning data are still current. Planning stresses the use of deception. Escape and evasion plans, external communication nets, and the location of caches must be known to leaders at all levels.

b. **Reverse planning sequence.** Insertion planning is done in the reverse planning sequence with the ground tactical plan being prepared first. Actions at the objective serve as the basis for the landing and assault plan, the plan for insertion and movement, the loading plan, and the plan for marshalling. All plans are interrelated. Some may be developed at the same time. Consideration of the factors of METT-T guides all planning. Special considerations that apply to these operations are:

(1) **Mission.** The nature of actions at the objective and requirements for special items of equipment and nonorganic personnel impacts on how the unit inserts. When the mission requires rapid deployment, the fastest method of insertion may be needed. In other operations, where success of the mission depends on maintaining secrecy, speed may be less important.

(2) **Enemy situation.** Enemy disposition, strength, and security measures along the route(s) near the objective area affect the method of insertion to be used. Planners must take into consideration whether the enemy:

(a) Uses RSTA detection devices or not.

(b) Has a strong air defense system or possesses air superiority.

(c) Has gaps between his defensive positions.

(3) **Terrain and weather.**

(a) **Topography.** Terrain such as mountains, swamps, or heavy forests favors land infiltration. Deserts, arctic regions, and open prairies favor air or water insertion. Areas not inhabited or with a friendly but scattered population favor land infiltration. A heavily populated area, or one which contains large enemy forces, favors air or water insertion. Hydrography--tides, currents, surf, reefs, and sandbars--must be considered when planning water insertion.

(b) **Weather.** Bad weather with reduced visibility and high winds favors land infiltration, and adversely affects air or water insertion. Limited visibility favors land infiltration; good visibility does not. Adverse weather aerial delivery systems (AWADS) lessens the impact of reduced visibility as a limiting factor on air infiltration. High winds may preclude airborne insertion.

(4) **Troops.** The number of men to be inserted and their level of training may be limiting factors. A need for special skills may call for use of nonranger unit personnel whose physical stamina and ability to perform many infiltration techniques may not equal that of the rangers. Whether supporting forces and special items of equipment are available must also be considered.

(5) **Time.** The main concern is whether the ranger operation is quick-response or deliberate-response. During a quick response mission, planning, preparation, and rehearsal time is short. The timing of the insertion is crucial with respect to weather, the enemy, and if an insertion craft is available.

4-12. Air insertion.

The most rapid form of insertion is by air. Rangers and equipment can be delivered by parachute (static-line or freefall technique), fixed-wing aircraft (airlanded), or helicopter (airlanded, parachute, or rappeling).

a. Suppression of enemy air defense along the insertion corridor may be necessary. The USAF will normally assume the major responsibility for suppressing enemy air defense artillery (ADA) capabilities that may interfere with the inserting forces. This is done by a variety of sophisticated electronic countermeasures applied against enemy equipment and by strikes against known enemy positions. Army fire support elements, special electronic mission aircraft (SEMA), and ground jammers may participate, as well as Navy aircraft, jammers, and gunfire.

(1) A main danger area is the forward edge of the battle area (FEBA) or frontier areas where

the enemy employs most of his sophisticated weapons systems and air defenses. Other dangerous areas are population or troop concentration areas behind enemy lines protected by point air defense systems.

(2) Since the majority of the enemy's detection devices and air defense weapons may be located near the point of entry, fire support, smoke screens (even at night), and suppressive measures may be critical. The use of special items of equipment may be needed to counter the enemy's RSTA effort.

(3) Long-range fires (artillery and naval gunfire) should be planned on known and suspected enemy antiaircraft locations and on prominent terrain features along the route(,s). Once beyond the FEBA, inserting units may be beyond the range of conventional indirect fire. More reliance will then be placed on air assets for fire support and electronic countermeasures.

b. There must be close coordination between the ranger unit and the transporting unit. What the transporting unit does during the insertion has a great impact on the success of the mission. To decrease the chance of detection, maximum use is made of periods of reduced visibility and tactical cover and deception. Drop zones and landing zones are chosen behind tree lines, in small clearings in forests, or on other types of terrain not easily seen.

c. All flight routes over enemy territory should be planned in order to avoid occupied areas and enemy air defenses. Routes should complement the cover and deception plan. They should avoid giving away final ranger objectives.

d. In-flight emergencies, mainly in deep-pedetration flights, should be considered, The ranger force receives a preflight briefing on the route to be flown and is informed of checkpoints along the route. Simple ground assembly plans for contingencies are set up before enplaning. In an emergency, the leader of the unit involved decides whether to continue the mission or abort. This is based on METT-T factors, contingency plans, and distance to the objective as compared to the distance to friendly territory. Contingency provisions for air and water rescue are made.

e. Deciding the DZ location is a joint effort. After receiving input from the air and ground commanders, the joint task force commander makes the final decision. During air assault operations, the ground commander is responsible for the overall process. After reviewing MEYF-T factors, it may be determined that surprise is more important than speed, or vice versa. This may determine the number, size, and locations of DZs and LZs.

f. Planning and preparation continue during the marshalling phase of any operation using air delivery for insertion. A review must be made of items of equipment that are crucial to the ranger force reaching its target area, accomplishing the mission, and returning safely. The need for special supplies and equipment, their transport, the distance to be traversed, the delivery means, and external support are all factors that enhance or reduce the chance for success. Once the means of insertion has been decided, logistics planning continues with emphasis on requesting and coordinating external support. The following must be considered:

(1) Detailed planning for mission logistical needs.

(2) Extra sets of key items of equipment to ensure redundancy and self-sufficiency of

inserting units.

(3) Plans for accompanying, planned, and emergency resupply. The ranger force should prepackage loads to meet specific contingencies. It should consider the use of supplies and equipment already in the operational area.

(4) In quick-response operations, only manportable items of equipment needed to accomplish the mission are carried. Plans and signals must be established for resupply of key items.

(5) Specific plans to evacuate casualties.

(6) All attached personnel and augmentation equipment must be properly prepared for movement.

(7) Provisions must be made for evacuating captured enemy equipment and personnel.

(8) All inserting units should be cross-loaded. Items to be considered in a cross-loading plan are the ground tactical plan, the assembly plan, the configuration of the DZ, the light route and formation of the aircraft, and the number of aircraft available. Cross-loading ensures that key leaders and equipment--for example, door bundles--are evenly distributed throughout the assaulting force. If one or more aircraft abort or are shot down, some key leaders and equipment will still arrive at the DZ. This permits mission accomplishment. Cross-loading aids rapid assembly on the DZ in support of the ground tactical plan. If a ranger is to assemble near the leading edge of a DZ, he is loaded on the aircraft to exit near the front of his stick.

(9) A bump plan is used if an aircraft should abort while still on the ground and there are no spare aircraft. The bump plan ensures that key leaders and equipment are not left behind when the main body takes off. If an aircraft aborts on the ground, key personnel move to another aircraft and replace nonessential personnel. This concept promotes success on the objective.

4-13. Airborne insertion.

a. During the planning phase of an airborne operation, there is no room for error in judgment. It can be planned and conducted well only if there is a constant exchange of information. All personnel involved must be informed of all changes in plans, times, and locations.

b. The reverse planning sequence is the key to the success of an airborne insertion. Most of the planning is based on the ground tactical plan, followed by the air movement plan to support it.

c. The USAF is responsible for flight planning, initial point (IP) selection, and crew procedures throughout the flight. The senior ranger in the aircraft must remain oriented and keep abreast of any last-minute changes. He must coordinate with the aircraft commander.

d. Most airborne insertions are made at either very high altitudes (HALO/HAHO) or very low. For combat operations, drop altitudes may be less than 500 feet above ground level (AGL), depending on the type of parachute used.

e. In airborne insertions, emphasis is on the use of special delivery or navigational techniques since normal insertion is during periods of limited visibility.

(1) **Adverse weather aerial delivery system.**

(a) A self-contained navigational system in some USAF C-130 aircraft is AWADS. The system allows the aircraft crew to fly a specific route to a DZ during darkness or bad weather. The system works with a Doppler radar that determines the ground speed of the aircraft, and an AWADS computer programmed with special flight route information (start point, route, checkpoints, wind speed and direction, and air speed). These two instruments show where the aircraft is in relation to the programmed checkpoints. This eliminates the need for airdrops under visual weather conditions.

(b) Only certain aircraft in formation are equipped with the actual AWADS device. Other aircraft are equipped with station-keeping equipment (SKE). This SKE consists of a computer that tells the aircrew where their aircraft is in relation to other aircraft in the formation.

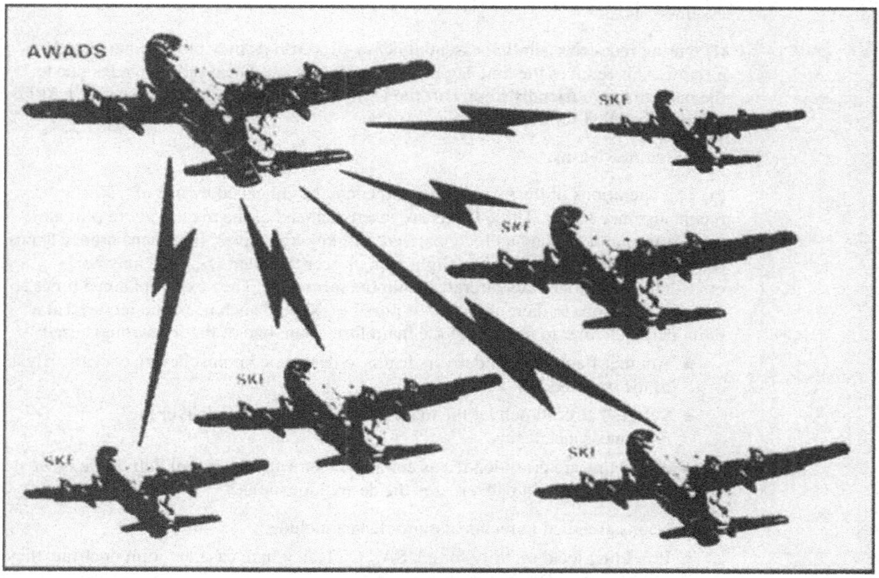

Figure 4-1. AWADS abd SKE aircraft formation.

(c) Airborne insertions using AWADS give the ranger commander the ability to airdrop personnel and equipment during low or zero visibility. This increases the safety of USAF aircraft. It protects them from visually tracked enemy antiaircraft guns and non-all-weather fighters. It also uses adverse weather as a combat force multiplier. Using AWADS, personnel and equipment may be inserted day or night without a pre-positioned USAF combat control team (CCT) or an Army assault team. Time length of the air train is shortened. The air corridor has to be cleared only once. This enhances the element of surprise. The CCTs are brought in with the assault elements. They are able to direct other low altitude parachute extraction systems (LAPES), container delivery systems (CDS), and air-land or heavy drop missions. Use of AWADS enables the ranger commander to conduct rapid, vertical reinforcement of units threatened by an enemy attack during bad weather.

(d) The use of AWADS requires detailed planning, rehearsal, and close coordination among the forces involved. Intense training of both USAF and Army personnel is required.

(e) Personnel must be briefed on terrain through use of photographs, maps, and terrain models. Limited visibility relies mainly on compasses and audible and visual assembly aids.

(f) During reduced visibility, essential heavy-drop serials may precede personnel aircraft. This reduces the time lag between personnel and heavy-drop chalks due to the need to move friendly troops off the DZ. Remaining heavy-drop serials or LAPES can be scheduled following P-hour.

(2) **Reconnaissance teams.**

(a) The operations of the main ranger force may be enhanced by use of reconnaissance teams. These teams are inserted ahead of the main body to provide early warning or to conduct reconnaissance of key objectives. If reconnaissance teams are required at a point along the flight path or near the main DZ, they may be cross-loaded on board any aircraft within the formation. They exit at planned times so as to land as close to their objective as possible. Should such teams be required at a point perpendicular to the axis of the main formation, one of the following is used:

- Another flank aircraft delivers teams to designated points before or with arrival of the main body.
- Selected aircraft deliver the main body, turn, and then deliver the reconnaissance teams.
- The teams are cross-loaded as complete elements on aircraft delivering other units whose flight path crosses the desired locations.

(b) Missions assigned to reconnaissance teams include:

- Providing local security for a USAF CCT, in which case, by joint doctrine, they form the Army assault team (AAT).
- Establishing surveillance or blocking positions on likely avenues of approach into the objective area or DZs.
- Locating and marking targets or assembly areas for the main assault force.

(c) Reconnaissance teams may be inserted by static line or HALO/HAHO operations.

(d) If the reconnaissance element is to locate a specific target or objective, it may be inserted up to 24 hours early. The longer the reconnaissance team remains near the objective area, the greater the danger of it being seen.

(e) In order for the information to be used by the ranger force commander, secure communications and a linkup plan must be prepared and rehearsed.

(3) **Assembly.**

(a) The ranger airborne force can conduct a quick assembly of combat power (weapons, personnel, and command and control elements) on the DZ. Quick assembly and reorganization is crucial because the unit is open to attack. Assembly must be fast and precise. Assembly areas and plans are set forth after considering all the factors of METT-T, mainly the location of the enemy, visibility, terrain, drop information, dispersion pattern, and cross-loading.

(b) The number of assembly areas depends on the location of the DZ, size of the assembly areas, and enemy's detection ability. Unit members move directly to the assembly areas without breaching security. Dissemination of information to all jumpers is the key to a successful assembly plan.

(c) For battalion-size or larger operations, general assembly areas are prescribed in the ranger force OPORD. Company commanders select and designate the exact location of their assembly areas. Although assembly area locations are terrain-dependent, they are generally on the DZ if the drop is at night and off the DZ if the drop is during daylight.

(d) Personnel chosen to carry assembly aids are cross-loaded in the center of their unit sticks. Strong, aggressive leaders should jump first and last in each stick. Upon landing, the leaders roll up their sticks toward the center, locate and move to the proper assembly aid. They may also carry or wear assembly aids. Individual rangers then follow these key personnel. During daylight operations, radiotelephone operators (RATELOS) can display color-coded pennants on their radio antennas to facilitate linkup with leaders. Key personnel may carry and jump with their own radios, and also display color-coded pennants on the radio antennas. Key personnel and RATELOs put their radios into operation before moving to assembly areas. For drops occurring near dawn or dusk, both daylight and night assembly plans are required.

(e) When tactical vehicles or door bundles are dropped, they are normally marked with panels, chemical lights, or strobe lights. The delivery parachutes are often color-coded.

(f) A ranger force should be able to assemble on an unfamiliar DZ, at night, within 45 minutes of the final aircraft pass. Because most ranger operations are time-sensitive, the unit normally moves from the assembly area to the objective when enough of the force has assembled to accomplish the mission. A straggler control point remains at the assembly area under the control of a designated leader. When everyone assembles, or at a specific time, the leader moves to link up with the unit. If available, a radio is left with the straggler control point.

(g) Assembly area security plans are prepared and disseminated. These plans must address hasty defense, sectors of responsibility, observation posts, fire support, and chain of command.

(4) **Contact on the DZ.** During the conduct of parachute insertion, rangers must be ready for enemy engagement at all times. Rangers are open to attack on the DZ before assembling. Just as contingency plans and chance immediate-action drills are developed, briefed, and rehearsed for other chance enemy contacts, rangers must be prepared to execute immediate-action drills for enemy contact on the DZ. The following actions are taken when there is enemy contact on the DZ:

(a) Rangers immediately assemble in fire team-size groups and fight off the DZ to designated assembly areas. Elements attempt to break contact and assemble.

(b) Assembly areas designated by commanders in the mission OPORD serve as primary rally points.

(c) Alternate assembly areas (normally 2 to 5 kilometers from the DZ) are designated in the mission operation order (OPORD). They are used by elements that cannot link up or assemble in the primary assembly area. These elements continue with the mission once assembled.

(d) During limited visibility, personnel should move directly and quickly off the DZ. They must avoid mistaking other friendly elements for enemy, which could cause firefights among friendly forces.

(5) **Reorganization.**

(a) The unit reorganizes according to plans, using designated assembly areas, aids, and identification markings for personnel and equipment. Replacements are named for key leaders or personnel who are lost or injured.

(b) Security teams assemble and move to their positions. Remaining units move quickly to their assembly areas, carrying supplies and equipment.

(c) Subordinate leaders keep the commander informed during assembly and reorganization.

(d) Runners may be used to report assembly status. This maintains radio silence. Designated personnel quickly recover supplies and equipment and move to the assembly areas.

(e) Reorganization is complete when all units have assembled or have been accounted for, and control has been established.

(6) **Pop-up technique.** This involves close terrain flying until the DZ is reached, then the aircraft quickly climbs to jump altitude. Special navigational equipment and aircrew training

are needed but may not always be available.

(7) **HALO/HAHO.** This involves high-altitude, low-opening and high-altitude, high-opening jumps with high performance parachutes designed to maneuver the jumpers to a specific point on the ground. During these operations, midair assembly procedures may be used.

(8) **Arctic.** Airborne insertion in extremely cold conditions is difficult. Times allocated for all parachute and equipment rigging must be increased. Rigging of individual equipment must be IAW FM 57-220. Ahkios can be rigged, packed inside a double A22 container, and dropped in platoon bundles. Because of the increased amount of individual equipment needed, allowable cargo loads for USAF aircraft are reduced.

(9) **Water.** Because of a lack of suitable land DZs, or to avoid detection, rangers may conduct airborne operations onto water DZs close to land. Procedures and techniques are similar to land DZ operations.

(10) **In-flight rigging.** On long-range operations involving extended flight time to the objective area, personnel may rig in flight. In-flight rigging requires detailed planning, coordination, of all jumpers. A rehearsal of these procedures should be conducted before deployment. Aircraft loads must be reduced to carry out in-flight rigging procedures. (See FM 57-220.)

4-14. Air-land insertion (fixed-wing).

a. **Planning factors.** Most of the planning factors that apply to airborne insertion operations also apply to air-land insertion operations. Most USAF aircraft cannot conduct air-land operations in total instrument weather conditions. The AWADS equipment is accurate enough for airdrop navigation, not air-land. Air-land operations are conducted, however, under conditions of extremely reduced visibility. Specially trained USAF crews, using sophisticated navigational techniques, can locate and land on small airfields in bad weather. As in airborne operations, ranger reconnaissance teams, as well as specially trained CCT elements, can jump before the air-land aircraft arrive. They emplace navigational and location markers to assist the pilots. Rangers may rely solely on air-land operations for insertion. This method does not provide for delivery of the force in the shortest time possible. Air-land insertions are normally in concert with an insertion of airborne elements.

b. **Time.** Since turnaround time is crucial, loads in aircraft should be unchained as the aircraft taxies. This also applies to personnel and seat belts. Aircraft should stop as close as possible to the assembly areas where personnel and equipment are off-loaded.

c. **Method.** Air-land insertions normally begin after a ranger Army assault team (AAT) and a USAF CCT have parachuted into the initial objective area and conducted reconnaissance, clearing, marking, and security operations. The ranger ground force commander is the arrival airfield commander. He has control over all arriving, taxiing, and departing aircraft (fixed-wing or rotary) through the CCT, which locates near him. The ranger force command and control element is usually on the first aircraft in and the last aircraft out.

d. **Security.** Air-land operations concentrate large numbers of men and equipment, which presents an open target to the enemy. The ranger ground force commander must ensure rapid off-loading, assembly, dispersion, and movement toward the objective.

e. **Communications.** The ranger ground force commander must have secure radio communications with the ground force, CCT, fire support elements, and higher headquarters. Wire should be laid to connect ground command and control locations.

f. **Operation duration.** Depending on the factors of METT-T, the aircraft may remain on the ground for extraction while the mission is being accomplished or take off at once. They may then orbit, pending recall, or return at a prearranged time.

g. **Extraction.** Extraction of the air-landed ranger force is normally planned and accomplished like an insertion, only in reverse order. A simple, yet effective, way must be found to see that no ranger personnel are left in the objective area.

4-15. Air assault insertion.

Ranger air assault (air-land, rappel, or combination) insertions normally take place during limited visibility. They quickly place the ranger force on, or close to, the objective. The ranger force (depending on the mission) normally plans on fighting when it hits the ground, completing the mission, and then being extracted. Air assault insertions need detailed planning, surprise, flexibility, speed, shock effect, concentration of combat power, and precise timing.

a. **Ground tactical plan.** The reverse planning process is important. The ground tactical plan, developed from the mission assessment, is the first planning area. All other planning begins from this point. The ranger commander and his staff decide stated and implied tasks, and request the proper helicopters for the mission.

b. **Fire support plan.** The fire support available to the ranger commander may be artillery, naval gunfire, mortars, attack helicopters, and USAF or USN tactical aircraft. Because of the timing of ranger air assault operations, preparatory fires are often not used. If they are used, enemy ADA sites normally have priority.

c. **Landing plan.** This plan puts the ranger force into the objective area at the time, place, and proper sequence to support the ground tactical plan. The ranger commander selects primary and alternate LZs and the priority for landing. The exact location of an LZ is determined after a detailed analysis of METT-T. Rehearsals are important. The ranger force must quickly assemble, reorganize, and accomplish the mission. The senior ranger on each aircraft ensures that the personnel on his aircraft offload at the correct LZ.

d. **Air movement plan.** The air movement plan provides for the coordinated movement of the ranger force into the zone of action in support of the 'landing plan. Flight routes, air movement tables, flight formation, altitude, and air speed should be considered. The ranger commander must have secure in-flight communications with his subordinate commanders.

e. **Loading plan.** This plan supports the air movement plan and provides for the pickup zone (PZ) setup and control. It sets priorities for the commander for loading personnel, equipment, supplies, and cross-loading. Air-loading table, aircraft loading formations, bump plan, and rehearsals should be considered.

f. **Staging plan.**

(1) This plan includes PZ selection and operation. The selection of PZs and LZs needs planning and coordination for best use of air assault assets. Site selection in or near the objective must be coordinated in person between the supported ranger commander and the aviation commander. The tactical situation is the key factor in PZ and LZ selection. Other factors are size of landing zones, surface conditions, ground slopes, approach and departure directions, prevailing winds, obstacles, communications, cross-loading, aircraft command and control, PZ and LZ markings, and rehearsals.

(2) The PZs are set up and secured about 15 minutes before touchdown of first aircraft. The fire support plan must support the other plans. All supporting fires must be coordinated with the air mission commander (AMC). Because of the depth of most objective locations, USAF tactical aircraft will be the main fire support method. Orbit locations for fire support aircraft must be planned.

g. **Other planning considerations.** These include escape and evasion, actions at the last LZ, assembly plan, downed aircraft procedures, control measures, forward area rearm and refuel point (FARP), weather delays, and deception plans.

4-16. Amphibious insertion.

Insertion by water may be by SCUBA, surface swimming, small boats, submarines, surface craft, helocasting, or a combination of these methods.

a. **Planning.** Planning for waterborne operations must be thorough. The plan is decided upon between the ranger commander, the transporting unit, and the joint headquarters directing the operation. Plans must be made for all types of enemy action and weather while on board the transporting craft. The transporting unit is given information on a need-to-know basis. Information that could endanger the operation may be withheld until the mission is under way. Initial planning includes:

(1) Time schedule of all events, from the beginning until the end of the joint operation, to be used as a planning guide. Timing for each event must be exact; the success of the operation depends on it. (See Figure 4-2)

Figure 4-2. Waterborne insertion.

(2) Embarkation site, where troops and equipment are loaded on the transporting craft.

(3) Drop site, where the ranger force will depart the primary craft and load into smaller boats.

(4) Handing site, where the unit will beach its boats, or land directly from amphibious craft.

(5) Loading, when loads are lashed and waterproofed IAW unit standing operating procedures (SOPs). Inspections are important. A bump plan for key personnel and equipment must be developed.

(6) Cross-loading, which will help preserve the ranger force. location of leaders, crew-served weapons, and needed equipment are stressed.

b. **Landing site.**

(1) **Beaches.** Beaches are checked to find one that allows an undetected approach. Landing sites that cannot be approached from different directions are avoided. Stretches of defended or patrolled coastline should be avoided. If sand beaches are used, tracks and other signs of

operation that may compromise the mission should be erased. Rural, isolated areas are best. The area behind the landing site should provide an assembly area and concealed routes for exit.

(2) **Barriers.** Reconnaissance may find that good landing sites are not suitable because of barriers between the site and the objective. Such barriers may be patrolled roads, large areas giving little or no concealment, enemy obstacles or defensive positions, areas populated with enemy sympathizers, or an area used for enemy troop maneuvers and training. Unless a means of avoiding such barriers is devised, landing sites should be limited to those without barriers.

(3) **Other factors.** Beach selection should also consider:

(a) The scheme of maneuver.

(b) Surprise.

(c) Enemy dispositions.

(d) Distance to the target.

(e) Landing and exit sites.

(f) Trafficability of the beach.

(g) Cover and concealment.

c. **Routes.** The route the transporting craft takes to the drop site should be planned to deceive the enemy. A major change in the route immediately after the unit's debarkation could compromise the mission. The route should be the same as a route that would be used in some other naval operation--for example, minelaying or sweeping, or a naval patrol. An alternate route must be planned.

d. **Tactical deception.** There should be other plans to deny the enemy knowledge of the insertion besides the water approach route plan. Plans may include electronic countermeasures (ECM), diversions, feints, ruses, or demonstrations.

e. **Actions at the drop site.** Some operations may permit landing directly from the transporting craft. If not, primary and alternate drop sites must be set. The drop site should be at least 1,500 meters offshore to cover any noise in the loading and launching. If the enemy has a surface radar capability, a drop site should be chosen to avoid detection. The drop site may have to be several miles offshore or may need the use of ECM. The manner in which men, small boats, and equipment are loaded must promote fast action at the drop site.

f. **Navigation.**

(1) Ship-to-shore navigation (to the landing site) may be done by dead reckoning. The course may also be maintained by compass navigation, reference to a shoreline silhouette, radar, or prearranged signals from the shore.

(2) Boat handlers, navigators, and other experts are relied on in various phases of the

transport operation.

(3) Lead reconnaissance teams or scout swimmer teams may be used to plant radar, radio, or infrared buoys, or distinct, shielded lights. Boats may be equipped with radar reflectors or beacons to permit accurate tracking.

g. **Actions on the beach.**

(1) Boats hold offshore while ranger scout swimmers move to a covered and concealed position. After reaching the beach, they make a short listening halt. They check the beach and the area immediately beyond the beach for enemy activity. Swimmers signal the force "all clear," mark the landing site, and provide local security.

(2) When the boats land, chosen personnel move directly into covered and concealed security positions. They prepare to defend the landing site.

(3) If the ranger unit is not carrying equipment, boats may be moved off the beach and buried or camouflaged. If heavy equipment makes it hard to move the boats, chosen personnel start unloading at once. The remainder of the force secures the landing site. Each boatload moves into position under the control of a designated leader. Teams are positioned with the best cover, concealment, and fields of fire available.

(4) The ranger force leader sets up a command post from which he can control the operation. He may establish secure communication with his headquarters.

(5) Boats may be deflated and buried or camouflaged near the landing site or away from it. If the boats are to be hidden near the landing site, a team may be chosen to dig holes or cut brush for camouflage along with the unloading of the boats. A small team may be chosen to sweep the beach to erase tracks and drag marks.

(6) The ranger force should be assembled and prepared to move within minutes after the last boat lands. Every ranger must be briefed on the unit's plan should enemy contact be made upon landing.

h. **Insertion by air from ship.**

(1) Helicopters from a ship can launch the ranger force from over the horizon and move at a low level to the LZ.

(2) Helicopters may be vectored from ships to a landing zone. The aircraft may return to the ship while the ranger force conducts the mission, they may orbit offshore, or they may stay at the LZ. Other aspects of landing and assembling are the same as discussed for helicopter insertions.

i. **Helocasting.** This form of insertion combines helicopters and small boats in the same operation. It is planned and conducted the same as air assault operations, except that the LZ is in the water. While helicopters are moving at low levels and low speeds, rangers drop small boats and jump into the water. They then assemble, climb into the boats, and move to shore. Ranger elements are usually assembled and loaded into the small boats within 20 minutes of the last ranger entering the water. This type of operation is normally conducted during limited visibility.

j. **Fire support.** Fire support may be ship-to-air, ship-to-shore, air-to-ground, or ground-to-ground. Specific missions include:

(1) Suppressing enemy air defense.

(2) Neutralizing enemy field artillery within range of boat or helicopter routes.

(3) Employing on-call fires against shoreline targets.

(4) Destroying enemy aircraft.

k. **Contingency planning.** The following must be planned for:

(1) Actions on enemy contact.

(a) En route.

(b) Helocast site.

(c) Flares.

(d) Aerial attack.

(e) Small arms fire.

(f) Indirect fire.

(2) Bump plan with priority to key leaders, crew-served weapons, and radios.

(3) Downed craft procedures.

(4) Escape and evasion.

(5) High surf.

(6) Adverse weather.

(7) Separation.

l. **Rehearsals.** The ranger force must rehearse all aspects of the amphibious insertion. This includes boat launching, paddling, beaching and assembly, boat commands, and capsize drills.

4-17. Land infiltration.

Land infiltration to the objective may be the best way to complete a mission. This is normal when the enemy has total air control or effective air defenses in the operational area to be penetrated, or when the objective is close to friendly territory. Rangers can infiltrate over any type of terrain, in any climate. However, thick forests, swamps, and broken or steep terrain offer the best chance for success.

a. **Planning.** There are many methods of infiltrating by land, and each has its good and bad points. Plans are designed to get the infiltrating force to the objective area without being seen. Plans should include:

(1) Selecting concealed routes for surprise and protection. Primary and alternate routes are based on detailed map and aerial photo study, ground reconnaissance, and data from agent reports.

(2) Avoiding obstacles, populated areas, silhouetting (when forced to cross obstacles or ridge lines), enemy positions, main avenues of approach, and movement along often used routes and trails.

(3) Selecting the time of infiltration to make use of reduced visibility, such as darkness, rain, snow, and fog, and periods of reduced enemy alertness, such as early morning, holidays, and times following protracted movement or combat. The time of the infiltration is important during critical phases (crossing borders or passing through enemy troop concentrations or populated areas).

(4) Using speed and stealth. All rangers must know routes, linkup points, alternate linkup points, time schedules, danger areas, and actions on enemy contact and at linkup points.

(5) Centralizing planning and decentralizing execution. Centralized coordination must ensure that units are acting IAW cover and deception plans. Control should be given to the leaders of the infiltrating groups. Within their designated zones, these leaders select their own routes and formations. Leaders control the rate of advance. When the enemy has electronic detection devices, radio communications must be kept to a minimum. Key leaders normally go with the group that is using the fastest lane or the lane least likely to be detected.

(6) Seeding infiltration routes to the assembly area with unattended ground sensors. These sensors can help the infiltrating force by reducing the chance of surprise contact during the infiltration, and by helping in controlling the forward progress of the units.

(7) Planning for fire support, which must include coverage of routes, danger areas, assembly areas, objectives, and PZs.

(8) Rehearsing actions at the objective, actions on enemy contact, actions at danger areas, and movement techniques.

b. **Movement.** Land infiltration is conducted using one of three techniques:

(1) **Movement in small groups along one axis.** All members of the force use the best route. Small groups are harder to detect, easier to control, and do not compromise the total force if detected. This technique may take too long and it needs an assembly area or linkup point. If the lead group is detected, other groups may be ambushed.

(2) **Movement in one group.** This technique has no reassembly problem. The same route is used by all, making navigation easier. A large force can fight out of a dangerous situation easier than a small one. It has no coordination problems between other infiltrating units. This form of movement is easier for the enemy to detect. If the force is detected, the mission may be endangered.

(3) **Movement in small groups along several routes at the same time.** This technique avoids putting the total force in danger, is less likely to be seen, forces the enemy to react in

many locations, and makes it harder for the enemy to determine the size of the force or its mission. Some groups travel over poorer routes and have problems of assembly, control, and support; the capture of personnel may endanger the mission. Regardless of which technique is used, ranger units normally move to reduce the chance of enemy contact.

c. **Assembly area.** Following movement into the general area of operations, infiltrating units converge on the assembly area or linkup point. Assembly may begin at once after the passage of lines, or it may take place at any time before reaching the objective. The infiltrating force must remain undetected before assembly. If an infiltrating unit is seen at this time, the enemy may learn the location of the assembly area, and the objective. It may take action to destroy the infiltrating force.

d. **Linkup point(s).**

(1) Selection characteristics include terrain that--

(a) Is free of enemy troops.

(b) Offers cover and concealment from enemy observation and patrols.

(c) Does not restrict movement of friendly units.

(d) Accommodates the unit.

(e) Is easily identified.

(f) Offers escape routes to alternate linkup points.

(g) Offers concealment from air attack.

(2) Action sequence for occupying the linkup point.

(a) Halt short of the point.

(b) Establish local security.

(c) Conduct a reconnaissance of the linkup point.

(d) Occupy the linkup point.

(e) Emplace observation posts and listening posts.

(f) Link up with other groups.

(g) Conduct final preparations.

(h) Continue the mission.

(3) Security may be enhanced with the aid of devices such as night vision goggles and night sights. Unattended ground sensors may be placed on likely avenues of approach into the linkup point. Normally, only passive observation devices are used to assist in avoiding detection.

(4) Plans must set a time schedule for actions at the linkup point. All unit members must move in quickly, consistent with security, so that the unit may proceed to the objective rally point (ORP) and on to the objective.

4-18. Stay-behind technique.

A ranger force can be positioned behind enemy lines by employing the stay-behind technique. In this method, a ranger force allows itself to be bypassed as the enemy advances. This is done to allow the ranger force to perform a specific mission. Stay-behind operations may require the concealment, or cache, of supplies before the enemy bypass. Rangers may need to use captured enemy supplies. The location of the assembly area and the plan for extraction are important parts of stay-behind plans.

4-19. Extraction.

The ranger force is normally extracted after the mission is completed. The enemy may also force the unit to abort its mission and to call for extraction. Contingency plans must be made to extract a force under enemy pressure, as well as one that is not in contact. Primary and alternate extraction sites are planned and coordinated with all ,porting forces. Ranger operations are normally conducted against deep targets. Extraction may begin with land movement but normally it includes the use of boats or aircraft. Plans may include primary I alternate means of extraction.

a. **Terrain.**

(1) The terrain is important in choosing the point of extraction - for example, whether or not the extraction site offers good cover from enemy direct-fire weapons and positions from which the extracting force can suppress the enemy. Tidal conditions and PZ size must be considered.

(2) Unlikely locations such as swamps, jungles, or mountain areas are often used for extraction. Specially trained air and naval crews, using the latest extraction devices, make such terrain useable.

b. **Enemy.**

(1) Extraction is best conducted without enemy pressure. However, such pressure could develop during the operation. Detailed plans are made for emergency or contingency extractions forced by the enemy as well as clandestine extractions not under enemy pressure. If under pressure, units will have to use planned fire support to suppress enemy weapons. At the beginning of the extraction with security posted to provide early warning, the unit may start to thin out the same as with a night withdrawal. Ground security may be relied on until the last unit withdraws under cover of suppressive fires.

(2) If enemy contact is made during an extraction, the commander must decide whether to reinsert units or break contact and move to another extraction point. Fire support and suppression are important and include suppressing enemy air defense by deception, ruses, or decoys.

(3) The ranger unit may have special weapons systems or equipment to aid in the extraction. Shoulder-fired antiaircraft missiles may be used to defend against enemy aircraft threatening

the extraction.

c. **Extraction by air or water.** Extraction by air or water is preferred when the resources are available and their use will not endanger the mission. Other factors that favor their use are when--

(1) Long distances must be covered.

(2) Time of return is essential.

(3) The enemy does not have air and naval superiority.

(4) Heavily populated hostile areas block land movement.

(5) Ranger forces are burdened with prisoners of war,casualties, or critical items of equipment.

d. **Extraction by land exfiltration.** Stealth and evasiveness are key elements of exfiltration. This method of extraction is favored when--

(1) The enemy has air superiority.

(2) The enemy is able to prohibit air or water extraction.

(3) The distance to friendly lines is short.

(4) The terrain provides cover and concealment (for movement on foot) and limits enemy mobile units.

(a) Multiple routes of exfiltration may be used to prevent destruction of the force if soldiers or units are detected. Captured enemy vehicles and equipment may be used to assist in the exfiltration.

(b) The ranger force may exfiltrate in one body or in small groups. Exfiltrating in small groups avoids the time delay of assembly.

(5) The exfiltrating force is lightly equipped and not burdened with captured personnel or material.

(6) The exfiltration route passes through an area occupied by friendly civilians or guerrilla forces who can assist the movement.

(7) Areas along exfiltration routes are not inhabited.

(8) The enemy force is dispersed or is not able to concentrate against the exfiltrating force.

4-20. Escape and evasion.

a. A vital part of all premission planning is the development of en-route and post mission E&E plans. Such plans enhance survival of the ranger force, and the transporting air or boat crews. The ranger force commander is responsible for--

(1) Helping in the development of an E&E plan in coordination with all supporting agencies.

(2) Ensuring that all members of the ranger force and the supporting boat or air crew are briefed on the E&E plan.

b. Each E&E plan is unique. Each situation has unique problems associated with E&E. The E&E plan devised by the ranger force commander must address these problems, while profiting from the individual abilities and training of the rangers and their supporting air and boat crews. The following considerations apply to E&E plans devised during ranger operations:

(1) The plans enhance survival of personnel who can no longer accomplish their assigned mission. The senior ranger in a group must decide if there are any remaining missions that the group can accomplish. If not, then he must try to evade and escape enemy capture if unable to link up and be extracted with the rest of the ranger force.

(2) Because of the depth of penetration behind enemy lines, most successful E&E plans may involve either air or water movement away from enemy-held territory. The air and boat crews supporting ranger operations are highly skilled and dedicated. If there is any chance of a successful rescue, these forces continue to try to retrieve rangers escaping from an enemy-held area.

(3) Rangers possess stamina and endurance. They can move a great distance (especially at night) and over rugged terrain to reach an area where they can rendezvous with search-and-rescue (SAR) aircraft or boats. Rangers are also resourceful and highly trained in communication, stealth, and camouflage. This, coupled with their self-discipline, means that successful E&E can be conducted over a great distance, through heavy concentrations of enemy, and over a long period.

(4) Escape and evasion plans for ranger elements should include avoiding contact with the local civilians. However, the aid of friendly insurgent forces may be enlisted. The ranger E&E plan may also include the use of other special operations forces E&E networks that are in place behind enemy lines. These nets must not be compromised by the volume of evaders.

c. Escape and evasion plans should be based on three phases:

PHASE ONE: That portion of the insertion following entry into the hostile zone where personnel survival and recovery is the only thing considered. Should an aircraft or boat carrying rangers be disabled during phase one, the ranking survivor assumes command. He must consider E&E planning and the experience and skills of other survivors. He should be concerned with the security of the mission, caring for wounded, and moving away from enemy activity while contacting SAR elements for recovery.

PHASE TWO: That portion of the insertion that is near the objective and could permit the ranger element to successfully pursue its mission. Although part of the E&E plan, actions by the ranger force during this phase become a forced land infiltration rather than an E&E. The senior surviving ranger must decide if enough personnel and equipment can be moved to the objective area in time to accomplish the mission. If so, he must proceed, taking along any surviving air or boat crew members. If not, he may proceed directly to the extraction point and wait, or he may contact SAR elements for recovery.

PHASE THREE: That portion of the operation after mission accomplishment. It is a hard time to escape and evade since the enemy is alert and reacting to the ranger forces. The confusion caused by the violence of the attack, coupled with the rangers' ability to move quickly cross-country away from the objective area, offers the best chance of a successful escape and evasion. The surviving ranger elements that cannot be extracted can do one of the following:

(1) Move over land at once to a PZ or beach for extraction by SAR elements.

(2) Move to an area where reentry into friendly lines can be made.

(3) Move to a neutral or nonaligned country where either internment or return to friendly control can be effected.

d. Any plan for E&E should provide for the issue to each ranger of an area map, a covert means of recognition, and a small compass. An emergency ration may also be included and any items needed for the climate in that objective area. The ranger force commander must balance the need for other survival equipment against the added weight and bulk involved.

CHAPTER 5

Strike Operations

The ranger regiment conducts strike operations in support of the Air Land Battle. Ranger strike operations are integrated into an overall plan designed to destroy, delay, and disorganize the enemy. They may also cause him to divert a major part of his combat forces and power to rear area security. Strike operations also help create an environment in which other special operations forces can operate. Strike operations can complement air power in disrupting enemy forces in depth. They allow the theater commander to regain the initiative and lead to decisive action.

5-1. Application.

a. The ranger regiment can conduct strike operations either as deliberate-response or quick-reaction missions. It can be inserted into the area of operations by land, sea, or air.

b. Strike operations can be conducted throughout the depth of the battlefield in support of conventional or unconventional warfare. Multiple strike operations can have a synergistic effect and can create situations that help friendly forces seize and maintain the initiative.

c. Strike operations include raids, personnel and equipment recovery, and interdiction of lines of communication. They are best accomplished by ranger units who use conventional techniques and special skills. Strike operations vary in size. When well augmented and supported, the ranger regiment can conduct strike operations deep in enemy territory using up to three battalions. Strike operations normally are limited in time and may be conducted unilaterally or in conjunction with other special operations forces or allied forces. They are normally ended by the extraction or exfiltration of the ranger force.

5-2. Raid operations.

a. A raid is a strike operation conducted behind enemy lines against strategic objectives, targets of high tactical value, time-sensitive targets, or key personnel and facilities in enemy rear areas. It normally involves a swift penetration of hostile territory to confuse the enemy or destroy his installations. Ranger units can conduct deep penetration raids when given the appropriate assets for insertion.

b. A strategic raid by the ranger regiment would be directed by the NCA and conducted under the OPCOM of a unified or JTF commander. An operational raid would be directed by the unified or JTF commander and conducted by that command directly or under the OPCOM of a corps commander. Typical targets would include:

(1) Command, control, communications, and intelligence centers at front level or combined arms army level.

(2) Nuclear and chemical weapons storage sites and delivery means.

(3) Key logistical centers, warehouses, ammunition complexes, or fuel pumping centers supplying logistical support to a specific front.

(4) Air defense and air traffic control integrating centers and air defense weapons sites.

(5) Key power generating and transmitting facilities, hydroelectric dams, and irrigation systems.

(6) Key ports or rail complexes.

(7) Key installations or facilities, such as airstrips, buildings, bridges, dams, or tunnels.

c. The size of the ranger raid force is determined by a detailed analysis of METT-T. The unit committed to the ground phase of the raid mission is kept as small as possible. It may be as small as a ranger company or it could include all the battalions of the ranger regiment. The force must be tailored to complete the mission quickly, violently, and with few casualties. The size of the force may also be affected by the augmentation of specially trained personnel needed for that mission.

d. Raids are normally conducted in five phases. (See Figure 5-1).

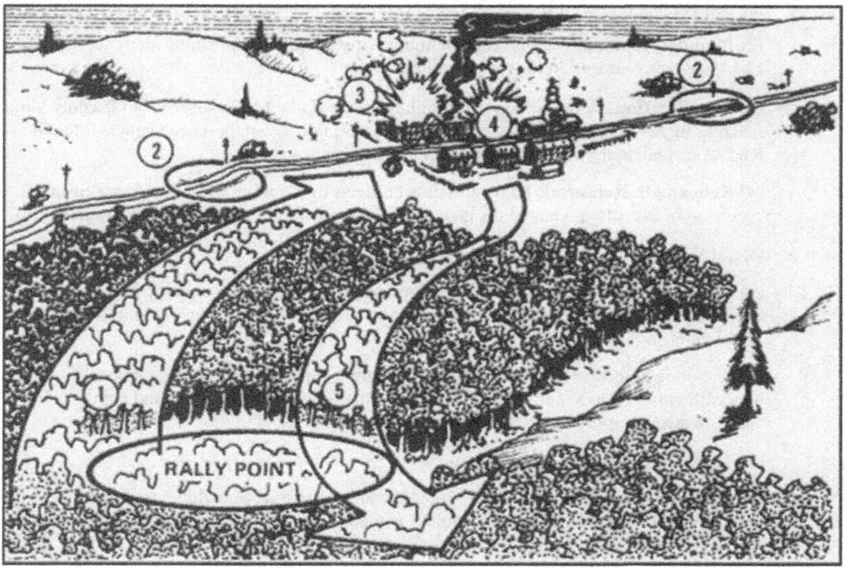

Figure 5-1. Five phases of a raid.

PHASE 1. The ranger force inserts or infiltrates into the objective area.

PHASE 2. The objective area is then sealed off from outside support or reinforcement, to include the enemy air threat.

PHASE 3. Any enemy force at or near the objective is overcome by surprise and violent attack, using all available firepower for shock effect.

PHASE 4. The mission is accomplished quickly before any surviving enemy can recover or be reinforced.

PHASE 5. The ranger force quickly withdraws from the objective area and is extracted.

e. This paragraph discusses the raid after insertion/infiltration. All considerations discussed in Chapter 4 must be applied in order for the unit to get deep behind enemy lines. Upon receipt of a raid mission, the following basics are emphasized:

(1) **Maximum use of intelligence information.** The gathering and dissemination of information must be continuous and provided to the raid force even while en route to the target area. To ensure mission accomplishment, the ranger unit must be kept informed of the latest enemy developments in the objective area to prevent being surprised. Only then can the full combat power of the rangers be concentrated at the decisive time and place.

(2) **Plan development.** The reverse planning sequence and the planning steps set forth in FM 101-5 are followed in the plan development.

(3) **Coordination.** Coordination is through the operational headquarters with friendly units (such as higher and supporting Army or joint headquarters, intelligence agencies) for fire, RSTA, special equipment, and personnel and logistical support.

(4) **Rehearsals.** Rehearsals help validate all aspects of planning for the raid and ensure precision in execution-They allow changes to be made in the plan before it is carried out.

f. A successful raid is ensured by--

(1) Launching the raid at an unexpected time or place by taking advantage of darkness and other periods of limited visibility, and moving over terrain that the enemy may consider impassable.

(2) Avoiding detection through proper movement techniques and skillful camouflage and concealment to include taking advantage of natural cover of the terrain, and using sophisticated equipment to detect and avoid enemy forces.

(3) Timing the operation as close as possible.

(4) Using all available support, both organic and nonorganic, to include use of special weapons such as Air Force "smart bombs" and artillery cannon-launched guided projectiles, with the ranger unit using laser target designators.

(5) Performing quick, violent, precise, and audacious actions that focus full combat power at the decisive time and place.

(6) Disengaging quickly upon mission completion.

(7) Withdrawing swiftly using planned routes and including a deception plan.

g. Four functions are normally performed by ranger tactical formations when conducting a raid. Each subelement is organized and equipped to do a specific part of the overall mission. Depending upon the specific mission, nature of the target, enemy situation, and terrain, these four functions are as follows:

(1) The command group controls movement to and actions at the objective. This unit normally consists of the ground commander, other subordinate leaders in the raid organization, and communications to support these leaders.

(2) The security element, whose organization is determined by the mission of the raid force, size and type of enemy force and its mobility and state of alert, terrain and avenues of approach into the area, and the time needed to seal off the objective area. The security element may perform the following:

(a) Secure the ORP.

(b) Give early warning of enemy approach.

(c) Block avenues of approach into the objective areas.

(d) Prevent enemy escape from the objective.

(e) Provide overmatch for the units at the objective and suppressive fires for their withdrawal.

(f) Provide short-range air defense.

(3) The support element provides the heavy volume of fire needed to neutralize the objective. Because fires from this unit are violent and devastating, they must be closely controlled to ensure the precision needed. On order or as planned, fires are lifted and shifted to cover the maneuver of the assault element by suppressing enemy fire from the objective or aerial fires. The support element may also be given specific locations to cover by fire in support of the security element if an enemy quick-reaction force moves toward the objective area. These may include routes into and out of the objective site, key terrain features, or installations adjacent to the main objective. Once the assault has been completed, or on order from the raid force commander, the support element displaces to the next planned position. Organization of the support element is determined by the following:

(a) Size of the objective, the geography of surrounding area, and the enemy threat (to include air) in the area. This element should be able to neutralize the objective (when supported by air or naval gunfire) and to lift or shift fires either when the assault is launched or when ordered by the raid force commander.

(b) Mission of the assault unit.

(e) Suitable firing positions.

(d) Size and nature of the enemy force in the objective area and those enemy forces capable of reinforcement at the objective.

(e) Fire support from other units (air strikes, naval gunfire, surface-to-surface missiles, and artillery fire).

(4) The assault element seizes and secures the objective and protects demolition teams, search teams, prisoner-of-way teams, or other teams.

(a) The organization of the assault element is always tailored to the mission. Each target area must be examined carefully. To assault, seize, and destroy an installation, position, or equipment, the assault element could be organized into one or more assault teams. The element's mission is to overcome resistance and secure the objective and to destroy the installation or equipment. Other specialized teams may also be needed. For example, sniper teams could be needed to remove key sentries. To capture prisoners, liberate personnel, and seize or destroy equipment, the assault element could be organized into assault teams, prisoner teams, search teams, medical teams, demolition teams, or breach teams.

(b) To destroy a point target or installation in a heavily defended area where the USAF cannot get close en to be effective, the assault element might be organized with one small team equipped with laser target designators. From covered and concealed positions, members of the team could then provide guidance for USAF delivery of laser-guided munitions from a safe distance.

h. The site chosen for the raid LZ or DZ must support the planned actions at the objective. There are two options when choosing sites:

(1) The ranger force can land on or near the objective and seize it before the enemy can react. (See Figure 5-2.) This avoids forced marches over land carrying heavy combat loads. If there is no suitable landing area near the objective, or the enemy has a strong reaction force nearby, this option is not favored.

Figure 5-2. Airborne landing on the objective.

(2) The ranger force can land unseen far from the objective. It then assembles, reorganizes, and moves into an ORP near the objective. (See Figure 5-3.) The objective is seized after security and support elements are in place. This option makes coordinated action easier by setting up control of small units before engaging the enemy.

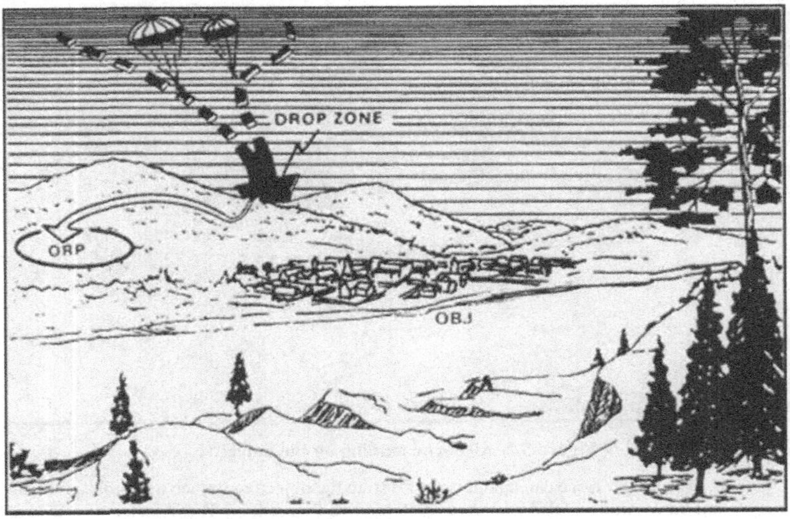

Figure 5-3. Airborne landing distant from the objective.

5-3. Interdiction operations.

a. AirLand Battle doctrine set forth in FM 100-5 is a shift in emphasis toward keeping the initiative in battle. The deep battle seeks out enemy elements separated from the main battle area and attacks them before they can be deployed. Interdiction of selected enemy forces by attacking them or blocking chokepoints hinders, delays, or interrupts the enemy's use of his lines of communication; it denies him the use of certain areas; it destroys industrial facilities, military installations, and equipment; and it stops or slows his advance along a particular route of approach. The ranger regiment's ability to conduct interdiction is vital to execution of the AirLand Battle doctrine.

b. Interdiction by ranger forces includes blocking or channeling approaches, holding key terrain for a limited time. and destroying facilities.

c. Ranger forces conduct interdiction missions that supplement those conducted by long-range Army and USAF weapon systems. Large-size ranger forces are normally used only when other systems fail or are not available, and the target is of such priority that it must be attacked. Small ranger laser target designator teams may work in concert with USAF or USN strike aircraft to direct terminally guided weapons systems. Rangers should be directed against crucial interdiction targets that disrupt, confuse, or delay the enemy and allow friendly forces to gain the initiative at the forward line of own troops (FLOT).

d. Typical interdiction operations are--

(1) Destroying rail lines, switches, rail yards, and rolling stock.

(2) Destroying bridges, overpasses, tunnels, and truck parks.

(3) Closing mountain passes or routes in restricted terrain.

(4) Cutting oil or natural gas pipelines and destroying pumping and distribution systems.

(5) Destroying dams, locks, spillways, and flood control gates.

(6) Destroying rail transload points, switches, and yard engines.

(7) Destroying shipyards, dry docks, wharfs, lighters, cranes, and stevedore facilities.

(8) Destroying electrical generation facilities, power lines, transformers, and grid monitoring centers.

(9) Destroying water pumping stations and filtration systems.

(10) Destroying radio and TV stations, phone lines, microwave terminals, satellite receiving stations, and telephone exchanges.

(11) Ambushing enemy forces and personnel moving along lines of communication.

e. Considerations when using rangers for interdiction are:

(1) Insertion and extraction.

(2) Support resources available.

(3) Ranger force vulnerability.

(4) Target identification.

(5) Threat environment.

(6) Other or supporting means of interdiction.

(7) Command and control.

(8) Time and magnitude of the effect on the enemy forces at the forward edge of the battle area.

(9) Collateral effects of target destruction.

f. Interdiction targets are not identified and attacked at random. They are part of an overall deep attack plan to destroy or cripple an entire system. Interdiction is based on the assigned deep campaign plan that directs which targets are to be attacked, by whom, the results desired, and the priorities of attack for specific systems. Based on his plan, the ranger force commander selects those ranger elements best suited to conduct the attack. Target selection, which requires detailed intelligence and planning, is based on:

(1) **Target criticality.** A target is critical when its destruction or damage would have a major influence on the enemy's ability to conduct or support operations. Each target is considered in relation to other elements of the particular target system designated for interdiction. The criticality of a target changes with the situation. For example, when the enemy has few locomotives, a railroad bridge is less critical.

(2) **Target accessibility.** The ranger force must be able to insert into the target area. Terrain, security controls, location, and means of insertion are considered when studying a target for accessibility. systems. Based on his plan, the ranger force commander selects those ranger elements best suited to

(3) **Target restorability.** The enemy's ability to restore a damaged facility to normal operating capacity affects target selection. The enemy's repair and maintenance ability and his reserve stocks must be considered.

(4) **Target vulnerability.** A target is vulnerable when it is open to attack by means available to ranger forces. Vulnerability is affected by the nature of the target such as type, size, disposition, and security. It is also affected by the means available to attack it such as explosives, incendiaries, and special devices.

g. Key to the interdiction process is the ability to locate and evaluate target systems for attack and the timing of the attack. A target system is a series of related functions that serve a common purpose.

(1) A target is one element, installation, or activity identified for attack such as a locomotive, train, bridge, or prison.

(2) A target complex is several targets in the same general area, such as a railway marshalling yard, an airfield, or port and dock facilities.

(3) A target system consists of an industrial system and its sources of raw materials; the rail, highway, waterway, or airway systems over which these materials are transported; the source of power and method of transmission; the factory complex; and the means by which the finished product is transported to the user. The targeting process is usually accomplished in the special operations force cell at EAC and the battle coordination element (BCE) at corps level.

h. Interdiction techniques include ambush, emplacement of conventional demolitions, target designation for laser-guided munitions, and emplacement of beacons or sensors.

NOTE: The enemy surface-to-air threat normally precludes the use of an airborne platform for direct observation and designation. This may be accomplished by the on-ground ranger force.

i. Through interdiction, ranger units are an ideal force to contribute to the Army's AirLand Battle operational concept. This includes taking the initiative, moving quickly to strike deep in the rear of the enemy, and defeating the enemy force beyond the FEBA. Planners and commanders must ensure that the ranger units are used on missions that are synchronized to contribute to the overall deep battle campaign. Strikes against critical targets must combine with efforts of other assets to take the initiative from the enemy, limiting his flexibility, initiative, and momentum.

5-4. Personnel and equipment recovery operations.

Recovery operations are specialized raids to liberate imprisoned or detained personnel or to return certain items of equipment to friendly control. These operations include recovering and extracting downed or hijacked aircrews and political or military leaders. Rangers use surprise and combat power to overwhelm resistance before harm can come to the detainees or prisoners.

a. **Planning.** The planning of personnel and equipment recovery operations is the same as for the raid. The main difference is that the ranger force commander must plan for the extraction of the recovered personnel and the loading and extraction of sensitive equipment. Personnel and equipment recovery raids are often executed as quick-response missions. They normally require the ranger force commander to plan and execute the recovery quickly.

b. **Augmentations.** This type of strike operation often requires that the ranger force be augmented with nonranger personnel with special skills or abilities. Examples are medical personnel, technical experts, mechanics or crew members trained in repair and retrieval operations, linguists or translators, and other special operations force elements.

c. **Organization.** The organization of the recovery force is the same as for a raid. Sometimes special teams are needed to perform certain missions involved in the recovery. This mission is conducted aggressively, usually ending in close combat during darkness or other limited visibility conditions. Reconnaissance teams may be inserted ahead of the ranger force to reconnoiter the objective and to locate the detained personnel.

(1) All planning and execution takes place as described for the raid. The emphasis must be on detailed, timely intelligence.

(2) Medical teams must be available to care for the detainees.

(3) The ranger personnel must be adept at searching and clearing of buildings.

(4) The ranger force must--

(a) Locate the specific building or area where the friendly detainees are being held.

(b) Gain entry into buildings, use stun grenades, and avoid obstacles and booby traps.

(c) Engage all enemy personnel using selective-kill techniques.

(d) Locate, identify, and secure detained personnel.

(e) Move personnel to a secured extraction site.

(f) Leave no detainees or rangers behind in the objective area.

d. **Execution.** The key to success is surprise and speed. The enemy must be assaulted when least expected. The ranger force must ensure that friendly detainees are not harmed during the assault - either by rangers or enemy personnel.

e. **Extraction.** The ranger force commander should plan for the extraction point to be as close as possible to the point of recovery. This prevents the detainees from having to move cross-country or for the ranger force to transport bulky equipment long distances. For small groups of detainees or small items of equipment, the extraction point can be further away. Recovery operations may use any method or combination of methods of insertion and extraction (see Chapter 4). Close planning and coordination is normally required with US Army, USAF, or USN aviation for both delivery and evacuation of the target area.

CHAPTER 6

Special Light Infantry Operations

Ranger units are normally employed on strike operations. However, broad US strategic needs demand a wide range of other military abilities. The ranger regiment may conduct special light infantry operations when larger conventional forces could not be used. The regiment conducts these operations using light infantry doctrine, methods, and techniques.

These operations include, but are not limited to, movement on foot, military operations on urbanized terrain (MOUT), passage of lines, ambush, and reconnaissance. Ranger units do not normally conduct planned operations against enemy armored forces. Battlefield surprises or METT-T considerations may cause ranger units to engage armored forces for short periods. The ranger regiment can conduct a conventional light infantry defense for short periods. A defense is not normally conducted adjacent to other friendly forces. It is normally a perimeter defense from which the unit then breaks out from encirclement and links up with friendly forces.

6-1. Movement.

a. Most missions involve movement on foot in tactical operations. Ranger units infiltrating behind enemy lines seek to avoid chance enemy contact. They choose the movement that allows them to retain security and control. To avoid loss of surprise and initiative, casualties, and mission failure, ranger units normally--

(1) Avoid chance enemy contact, if possible.

(2) Move on covered and concealed routes.

(3) Avoid likely ambush sites and other danger areas.

(4) Practice camouflage, noise, and light discipline.

(5) Maintain all-round security.

(6) Make unavoidable enemy contact with the smallest element possible.

(7) Retain the initiative to attack at the time and place of the ranger unit's choice.

b. In planning the movement of his unit, the ranger commander must consider all the elements of METT-T and determine how they will affect the movement. He must determine the route, formations, security measures, methods of command and control, location of key leaders and weapons, and action to be taken on chance enemy contact.

(1) **Routes.** The ranger force commander conducts a careful route analysis, using the factors of observation, concealment, obstacles, key terrain, and avenues of approach (OCOKA). He chooses a route that maintains the security of the force while ensuring surprise. He tries to choose a route that will avoid enemy contact and speed movement. The commander makes a detailed route reconnaissance using maps, aerial photographs, and reports from persons familiar with the area. If possible, he reconnoiters the route himself or uses a reconnaissance team. He may have that team act as guides or mark the route using covert marking devices.

(2) **Formations.** Ranger units use the movement techniques found in current light infantry doctrinal publications. The traveling, traveling overmatch, or bounding overmatch technique is chosen based on the likelihood of enemy contact. The ranger force normally tries to avoid detection and contact with the enemy while moving. The commander chooses a movement technique that ensures that if unexpected contact were made, it would be with the smallest element possible and not the entire formation. The distance between moving ranger elements depends on the terrain, visibility, and enemy situation. March interval is normally great enough to allow each succeeding element either to deploy or change direction if enemy contact is made. Distances are close enough that each element could quickly assist the element to its front.

(3) **Security.** Security is to the front, rear, and flanks when enemy contact is likely. A front security element is placed well forward with adequate communications with the main body. The security element is used to detect the enemy and warn the main body. The main body then either changes direction or hides until the enemy is past. If the enemy does detect the main body, the security element may be used to attack the enemy from the flanks or rear. Flank security elements are placed next to the route of march. They move forward either by alternate or successive bounds. An alternative is to have the security element moving adjacent to the column along routes paralleling the direction of march. Rear security is handled the same as front security. Plans are made for the rear guard to support the main body during chance enemy contact. This is done either by maneuver or by furnishing supporting fire.

(4) **Command, control, and communications.** All available communication is used, consistent with OPSEC and movement security, to assist in maintaining control during movement. March objectives, checkpoints, and phase lines may be used to aid in control. The number of reports is reduced, normally only exception reports are needed. The ranger force commander should be well forward in the formation but may move throughout as the situation demands. Communications with security elements are mandatory. Operations security often precludes the use of radios. Therefore, connecting files, runners, and visual signals are normally used. Detailed planning, briefing, rehearsals, and control are valuable if there is enemy contact. Alternate plans are made to cover all possible situations.

(5) **Weapons.** Rangers are placed within the formation to allow a mix of their individual weapons to cover the front, flanks, and rear of the formation. Rifles, grenade launchers, and light automatic weapons cover all sectors. Crew-served weapons such as machine guns, antitank weapons, and mortars should be given designated locations within the formation. This allows the element leader to employ them quickly. They should be placed near the lead element, but not so close as to be engaged along with it. Crew-served weapons are used to

provide overmatch, especially when employing the bounding overmatch technique of movement.

(6) **Rehearsals.** The ranger force, expecting movement as part of an operation, conducts the following combat drill rehearsals:

(a) Conduct movement.

(b) React to enemy contact.

(c) Maneuver under fire (fire and movement).

(d) Cross danger areas.

(e) Break contact.

(f) Conduct a hasty ambush.

(g) Make a temporary halt.

(h) Establish a perimeter.

(i) Establish a rally point.

(j) Take actions upon consolidation.

(k) Use arm-and-hand signals.

(l) React to flares and indirect fire.

c. After a ranger force is inside enemy territory, it must be alert to avoid detection before reaching the objective.

(1) Should the ranger force security element become aware of enemy presence, it alerts the main force. The main force tries to move away without alerting the enemy and continues the mission.

(2) If the enemy does make contact, the security element then engages the enemy, trying to deceive him as to the size of the ranger force and its direction of movement. After breaking contact, the security force divides into small groups and moves to a rally point to rejoin the main force.

(3) The ranger force does not fight en route to the objective unless there is no alternative, then it quickly breaks contact and moves away.

(4) Ranger units may need to set up a temporary defensive position for resupply, evacuation of wounded, planning, or extraction. One technique that can be used is the temporary perimeter defense.

(5) If the moving ranger force comes under indirect fire, it moves quickly out of the impact area and continues the mission.

(6) The most effective means of combating an ambush is through the immediate return of a

high volume of fire by all weapons. This requires alertness, discipline, dynamic leadership, and rehearsed plans.

(a) Regardless of the method of movement, all weapons should be positioned for immediate use. A well-executed ambush is hard to counteract. However, the effectiveness of an ambush can be reduced by the immediate return of fire. Enemy weapons may be silenced, thus creating weak points in the ambush position. Smoke and high-explosive (HE) grenades are effective in executing counterambush actions. They create confusion, provide a screen for movement, and disrupt the ambush plan.

(b) When all elements of a ranger force are not trapped within the ambush, the elements that are free to maneuver should initiate an immediate flank or rear assault against the ambush force. A flank assault permits better coordination with personnel in the ambush. Supporting fires are provided by those weapons that cannot be hand carried and fired in the assault.

(c) Specific actions against a near and far ambush are as follows:

- **Near ambush.** Elements caught in the kill zone return fire at once and move out of the kill zone. Elements not in the kill zone help by maneuvering and conducting a flanking attack. The unit assaults the ambush position.
- **Far ambush.** Those ranger elements not in the kill zone set up an overmatch position and lay down a base of fire for the elements in contact. The elements in contact move out of the kill zone and prepare to assault the enemy, using individual movement techniques and fire and movement.

(7) The reorganization after enemy contact involves the use of rally points, plans for local security, reorganization of the force, evacuation of dead and wounded, and movement based on the unit mission.

6-2. Ambush operations.

a. **Definition.** An ambush is a surprise attack, from a concealed position. It is used on a moving or temporarily halted enemy to destroy or capture him and his equipment. An ambush is a useful tactic because--

(1) Small, well-trained, disciplined forces, with limited weapons and equipment, can destroy much larger enemy forces.

(2) It reduces the enemy's overall combat effectiveness by destruction and harassment of his forces.

(3) Enemy morale and effectiveness suffer heavily at little cost to the force executing the ambush.

b. **Execution.** A successful ambush must be executed with precision, violence, speed, and audacity of execution. For success, ambush operations must emphasize the following:

(1) **Surprise.** Surprise, more than any other single aspect, enhances the value of an ambush. Surprise increases the potential for inflicting damage on the enemy with less danger to the

ranger force.

(2) **Coordinated firepower and shock effect.**

(a) Coordinated firepower is used for maximum shock effect. Massive volumes of accurate fire, explosives, and mines, coupled with an aggressive attack, breaks the enemy's spirit to fight back. Surprise increases shock effect and the chances for success. Shock effect can cover unexpected defects in an ambush--for example, ambushing a much larger force than expected.

(b) All weapons must be sited with interlocking fires in the kill zone and along likely avenues of entrance or exit. Mortars should be used if the terrain permits. Tripods and traversing and elevating mechanisms are normally used with machine guns to lock in fires. All riflemen use firing stakes to mark left and right limits, and elevation stakes. There is a tendency to shoot high in an ambush--especially at night. The M203 grenade launchers are sited to cover the dead space and routes of escape. When the rules of engagement permit, plastic bags containing CS crystals may be taped to the front side of Claymore mines.

(3) **Control.** Control is essential.

(a) Leaders must have contact with all members of their unit to alert them to the oncoming enemy. Leaders should not move around the ambush site during this crucial period. A method used to alert members can be to tie strings or vines to rangers' legs or arms. By a series of light tugs, all members of the ambush can be alerted to enemy presence.

(b) The leader must initiate the ambush with a casualty-producing device. A bank of Claymore mines on a double-ring main is an excellent device to spring an ambush. Other good techniques are to use a 90-mm recoilless rifle firing antipersonnel (APERS) rounds or a machine gun. All must be under the direct control of the ranger leader. Whistles or pyrotechnics must not be used. They will allow the enemy time to react.

- As soon as the enemy is hit, he reacts. The ambush force has only a few seconds to destroy the enemy in the kill zone. He will quickly recover from the initial shock and leave the kill zone--either with a direct counterattack or withdrawal, Subsequent fires and other banks of Claymore mines must be planned.
- The leader always executes the ambush, except when a member of the ambush knows he has been discovered. He then has the authority to execute--with killing fire, not by yelling.
- The cease-fire must be controlled by the ranger leader. A whistle or other device may be used to get attention, and then cease-fire is signaled.

(4) **Security.** The flanks and rear of an ambush site are open to counterattack. Flank and rear security may be enhanced by -

(a) Echeloning in depth.

(b) Designating sectors of observation.

(c) Positioning of RSTA devices.

(d) Enforcing noise and light discipline.

(e) Having a good withdrawal plan.

(f) Securing routes of withdrawal.

(g) Executing with speed and violence.

(h) Positioning security forces to sea] off the ambush area.

(i) Having good camouflage.

(5) **Simplicity.** A simple, direct plan improves the chance of success. The ambush plan must be clear yet concise to offer the greatest likelihood of success. For example:

(a) Mission statements for security, support, and assault elements must be clear, concise, and direct.

(b) Tasks to be performed by the ambush elements should be easy to understand.

(c) Contingency plans should be simple.

(d) Routes into positions and withdrawal routes should not cross. They should be the shortest, most secure routes.

(6) **Training and self-discipline.** Precise and violent execution is how ranger units win against enemy forces. All advantages must be exploited. Discipline must be strict. There must be no sleeping, talking, eating, or smoking in the ambush site. If an ambush is to be set up for long periods, then the elements of the ambush must be pulled back at set times to the ORP for rest. Extended ambushes of 24, 36, or 48 hours require six- or eight-hour shifts. It may take a company to man an extended platoon ambush position. Tired troops cannot man an ambush well. Other vigorous operations cannot be performed all day and have an alert ambush all night.

(7) **Maximum use of specialized equipment.** This includes the most sophisticated weapons, communications equipment, and RSTA devices. The effectiveness of any mission is enhanced by the use of this equipment.

(8) **METT-T.** Every ambush mission must be evaluated by the ranger commander IAW METT-T to design an ambush that works.

(a) **Mission.** Clearly state the mission--for example, stop enemy resupply, ambush enemy armor columns, or capture personnel or equipment.

(b) **Enemy.** State the type of enemy units--for example, guerrillas, main force, logistics carrying parties, or armor units. What is the size of the force, its habits when attacked, and type of support? What is the enemy night vision ability and expected time of movement?

(c) **Terrain.** Is it wooded, mountain, jungle, or urban? Some good ambushes can be set in a city or suburb. Are the routes available to the enemy suitable to his activity?

(d) **Troops.** Squads and platoons are the basic ranger ambush forces. An ambushing force should be able to engage a force two to three times its size--but evaluate enemy and terrain, A platoon ambushing an enemy tank company in daylight in open terrain is risky. But engaging that same column at night or in a forest may give the ranger platoon an advantage.

(e) **Time.** How long will it take to emplace the ambush? A complex, lethal ambush with plenty of "dirty tricks" takes hours to set up; a hasty ambush on a trail takes 15 to 30 minutes. How long is the ambush to be in place? What hours of the day, or at nighttime only? How soon must it be set up? Do not set up an ambush in darkness--even if it is to be executed in darkness. Try to allow every ranger to '6see" and "lay" in his weapon in daylight. Time is important.

c. **Organization of the ambush force.** A ranger unit conducting an ambush must be task-organized to perform the following functions: assault, security, and support. The ambush forces should be task-organized according to the TOE--by platoons, squads, and fire teams. The TOE should not be changed to create smaller elements for an ambush. The TOE formations may be reinforced with machine gun, recoilless rifle teams, or a 60-mm mortar squad.

(1) **Assault.** The elements assigned the assault mission either move directly into their positions or through a release point. The mission may include any combination of the following:

(a) Conduct the main assault.

(b) Halt an enemy's motorized column or any moving target,

(e) Kill or capture personnel.

(d) Recover supplies and equipment.

(e) Destroy vehicles and supplies.

Search teams are not always used. The ranger leader must decide how and when to use search teams. When rangers leave the security of their well-chosen, concealed ambush position, they are subject to the fires of the enemy who may also be hidden and ready. If it is at night, do not throw tripflares or shoot illumination to light the search area as this will illuminate and expose rangers to the enemy. Always assume there is hidden enemy--the ambush will not kill them all. Night observation devices (NODs) or a red filtered flashlight taped to M16s should be used to quickly search in ranger buddy teams. If the return fire from the enemy is great or if the ambush missed the main body, then the leader may choose to break contact and leave without searching the kill zone.

(2) **Security.** The elements assigned the mission of security may move to their positions

directly or by way of a release point. Their missions may include any or all of the following:

(a) Secure flanks, rear, or ORP.

(b) Provide early warning.

(c) Seal off the kill zone to prevent the enemy from escaping or reinforcing.

(d) Assist in executing the ambush.

(e) Cover withdrawal of main ambush force.

(3) **Support.** The units assigned a support mission provide fires that may include employment of--

- Heavy automatic weapon fires.
- Antitank fires.
- Mortar fires.
- Mines.
- Flame munitions.

d. **Ambush site.**

(1) **Choose an ambush site.** When choosing an ambush site, all sources of information must be used to enhance surprise, exploit the enemy's weak points, and take advantage of the terrain. An analysis of the terrain is also based on the other factors of METT-T to include maps, aerial photos, and, if possible, personal reconnaissance. Emphasis is on--

(a) Natural cover and concealment for the ambush force.

(b) Routes of entry and withdrawal (at least two) that should be direct and easy to reach.

(c) Good observation and fields of fire.

(d) Harmless-looking terrain.

(e) Few enemy escape routes.

(f) Limited enemy reinforcement ability.

(g) Nearby assembly or rendezvous area.

(h) Terrain that will canalize enemy into kill zones, and natural obstacles to keep him there.

NOTE: Try to select a site covered by friendly supporting indirect fires.

(2) **Take advantage of the terrain.** Emphasis must be on exploiting all natural cover and concealment afforded by the terrain. Site the ambush and individual positions based on the terrain rather than trying to adapt the terrain to a fixed geometric design.

(3) **Restrict enemy movement.** Restricting enemy movement by natural or man-made

obstacles should also be planned.

e. **Types of ambushes.**

(1) Ambushes have two categories for ranger purposes--area ambush and point ambush.

(a) Area ambushes may be set up by ranger platoons, companies, or battalions. They are used to interdict enemy movement in a given area or inflict casualties on his forces. Area ambushes consist of a series of point ambushes. The size and location of the ambushes are dictated by the METT-T analysis.

(b) Point ambushes are set at the most ideal location to inflict damage on the enemy. Such ambushes must be able to accept the enemy force from more than one direction. The ranger force levels may enable it to execute an ambush in two or three main directions. The other directions must be covered by security that gives early warning of enemy attack.

(2) The basic ambush is linear (Figure 6-1). An ambush is set on an expected avenue of approach. This ambush can accept contact from three basic directions--left, right, and front. The rear is secured by a security team.

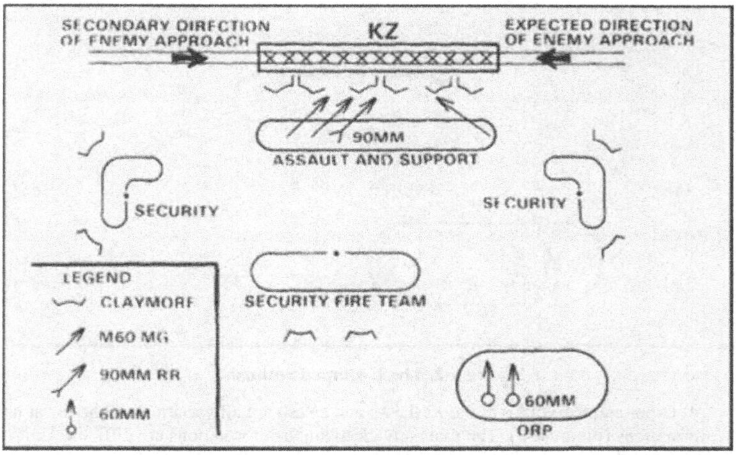

Figure 6-1. The linear ambush.

(3) A variation of the linear ambush is the L-shaped ambush (Figure 6-2). The L-shaped ambush is formed with the base (bottom) of the L perpendicular to the expected enemy direction of advance. This is a good ambush for a road, jungle trail, or an area where the enemy is canalized and his approach route is known. The L-shaped ambush can handle an enemy coming from the expected direction (toward the base of the L) from the front (stem

of the L). It is less effective against an enemy formation that comes from the opposite direction. The ambush must be executed when the enemy main body has cleared the base of the L. The flank security must protect the rear of the base of the L. In the L-shaped ambush, weapons must be sited to avoid direct or ricochet fire into friendly forces. The security elements must be considered as fire fans are being planned.

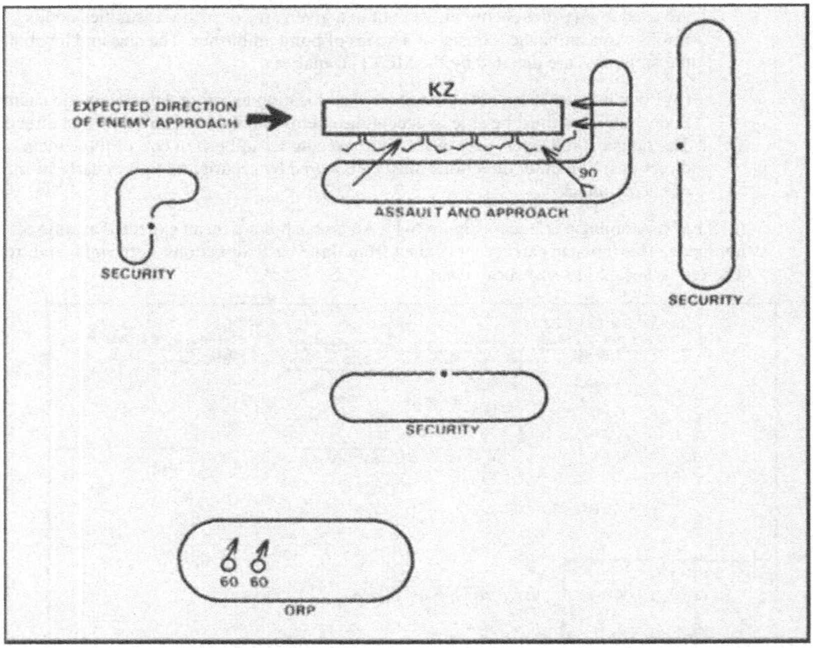

Figure 6-2. The L-shaped ambush.

(4) Other ambushes (the Z, the X, the V, and so on) are all variations of the linear and L formations (Figure 6-3). The more advanced ambush formations are difficult. Units must be expert in the basics before they are ready to employ such techniques.

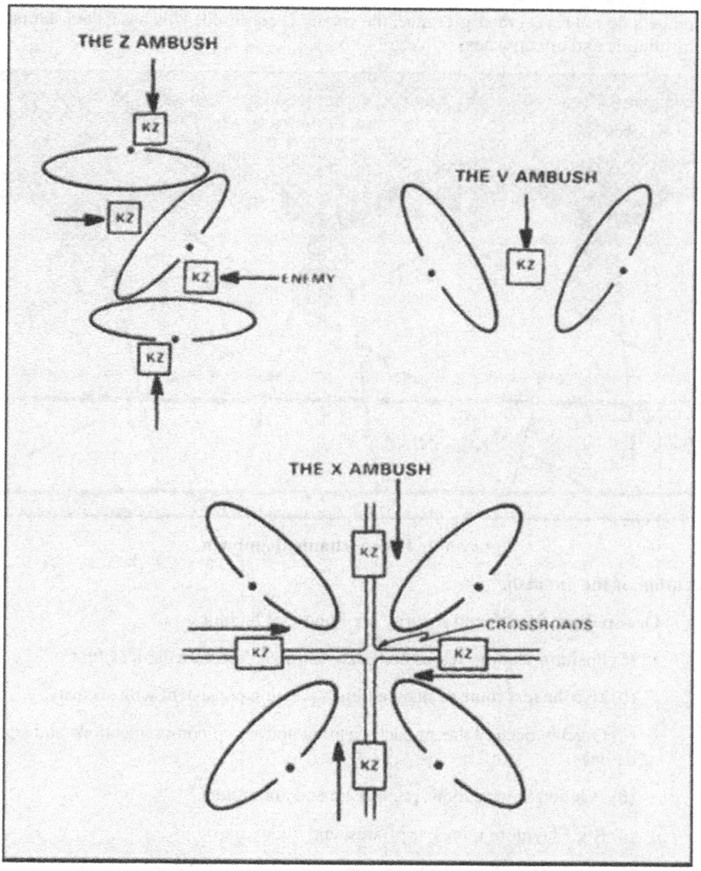

Figure 6-3. Other ambush variations.

(5) Another type of point ambush is the mechanical ambush (Figure 6-4). It consists of Claymore mines set in series with a double-ring main. It is command detonated or detonated by a triggering device activated by the enemy. Mechanical ambushes are normally manned. Rangers prepare to engage the enemy with direct fire if the mechanical ambush does not detonate or if it is wholly or partially ineffective. Mechanical ambushes are an effective way to interdict a large area using a small ranger force. If the mechanical ambush is effective and

rangers do not reveal their presence, the enemy is confused. This has a devastating effect on his morale and effectiveness.

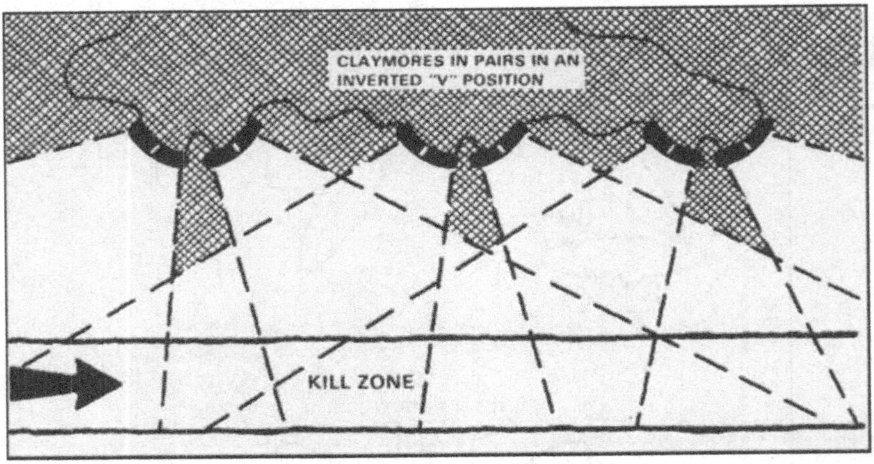

Figure 6-4. The mechanical ambush.

f. **Execution of the ambush.**

(1) **Occupation.** Stealth and security are important factors.

(a) Position security teams and early warning detection devices first.

(b) Use the rest route to main ambush position consistent with security.

(c) Quickly occupy the ambush position and set up communications and signaling devices.

(d) Position key weapons (automatic and antiarmor).

(e) Rig Claymore mines, tripflares, and booby traps.

(f) Ensure that all weapons are correctly positioned. Assign sectors of fire to provide mutual support and cover dead space.

(2) **Camouflage.** Each man must be hidden from the enemy, During preparation for the mission, each man camouflages himself and his equipment, and secures his equipment to prevent noise. At the ambush site, positions are prepared with minimum change in the natural appearance of the site. All debris resulting from preparation of positions is concealed to prevent evidence of occupation.

(3) **Movement.** Movement is kept to a minimum. The number of men moving at a time is closely controlled. Every man is as quiet as possible, especially at night. Light discipline is rigidly enforced at night. Smoking is forbidden.

(4) **Signals.** Audible and visual signals, such as whistles or pyrotechnics, must be changed often to avoid setting patterns and alerting the enemy. Three or four simple signals are needed to execute the ambush. Signals are used to--

(a) Provide early warning of an enemy approach. A signal by the security force to alert the patrol leader to the correct direction of enemy approach may be given. This includes arm-and-band signals, radio, or field telephone.

(b) Initiate the ambush. This may be the detonation of mines or explosives, Fire is then delivered at once in the heaviest, most accurate volume possible. Properly timed and delivered fires add to the achievement of surprise, as well as to the destruction of the target.

(c) Lift or shift fires, Signal for lift or shift fires, if the kill zone is to be assaulted. Voice commands, whistles, or pyrotechnics may be used. When the kill zone is assaulted, the lifting or shifting of fires must be as precise as when starting the ambush. Otherwise, the assault is delayed and the enemy has a chance to recover and react.

(d) Withdraw. A signal for withdrawal can be voice commands, whistles, or pyrotechnics.

(5) **Objective rally point.** The ORP is located far enough from the ambush site so that it will not be overrun if the enemy manages to attack the ambush unit(s). Existence loads are normally left in the ORP. Routes of withdrawal to the ORP are scouted (when possible by each man), Withdrawal routes should provide cover and concealment for the unit and hinder enemy pursuit. Withdrawal routes are a main consideration in the selection of the ambush site. They may be the key to survival after execution of the ambush. On signal, the ambush force quickly (but quietly) withdraws to the ORP. If pursued, the withdrawal may be by bounds, with mines or hasty ambushes used to delay pursuing forces.

(6) **Ambush variety.** If one method of ambush is used predominantly, the enemy will develop an effective defense against it and will be affected less by the shock of the ambush since he knows what to expect. More than one method of ambush should be used. No single method will fit all combinations of terrain, equipment, weather, and enemy capabilities. Variety should also be used when signaling - for example, use multiple signals, both audible and visual. Use weapons fire, mines, and RSTA when possible and vary signals to avoid compromise.

(7) **Swift action.** Speed in the execution of the ambush and the withdrawal should prevent enemy reaction forces from engaging the ambush force. Speed is often a shield against casualties and failure. When there is contact with reaction forces, speed may enhance quick disengagement. Ranger units must not only perform better than the enemy but also move faster than the enemy can react.

g. **Successful ambush.** Emphasize the following:

(1) Intelligence to ensure the enemy is ambushed at a time and place when he least expects or is least prepared to fight. This is preferred during reduced visibility to achieve surprise.

(2) Detailed planning and thorough training and rehearsing of all elements in all phases of the ambush. This ensures maximum shock effect through swift, precise execution.

(3) All available RSTA devices to permit daytime effectiveness at night when moving, shooting, or detecting enemy movement. To avoid detection, active RSTA devices should not be used until after the ambush has been triggered.

(4) All available firepower with emphasis on antiarmor, area and automatic weapons, and precision-guided munitions.

(5) Speed to achieve surprise and enhance security of the force.

(6) Cover, concealment, and overall protection afforded by the terrain when moving or when occupying ambush positions.

6-3. Antiarmor operations.

a. Antiarmor operations are not primary missions for ranger forces. They lack antiarmor firepower and have limited antiarmor ability. Ranger units can, however, perform limited antiarmor missions on close terrain by interdiction or ambush. The antiarmor weapons they do have are man-portable, but the ammunition is too heavy to carry many rounds cross-country. A ranger battalion has ten Dragon missile launchers and nine 90-mm recoilless rifles. It is also equipped with light antitank weapons (LAWs) and laser target designators (LTDs) that can guide Copperhead rounds.

b. Every ranger must be skilled with the limited antiarmor systems available. Proper selection, fortification, and camouflage of antiarmor ambush sites, route selections, withdrawal procedures, and initiative are important. Armor kill zones (AKZs) are normally set up after infiltrating the ranger force behind enemy lines under limited visibility.

c. A ranger antiarmor force is organized into four elements: command and control, support, security, and armor-killer teams (Dragons, LAWs, 90-mm recoilless rifles). These elements function the same as a raid force. The exception is that the assault force is the armor-killer team with the mission of destroying enemy armor, and it will probably not enter the AKZ. After the ambush occurs, the ranger force withdraws to its ORP. Indirect fires (if available) are then brought to bear on the AKZ.

d. Ranger units may conduct antiarmor ambushes and interdiction operation by use of laser-guided Copperhead rounds or air-delivered precision-guided munitions. Air-delivered family of scatterable mines (FASCAM) can be used to provide the ranger force security against attack by armored forces.

6-4. Military operations on urbanized terrain.

a. The ranger regiment normally tries to avoid the manpower-intensive, high-casualty battles associated with MOUT. However, due to the nature of strike operations and the types of targets selected, the regiment may conduct MOUT for limited periods. As in all ranger operations, the ranger force commander tries to surprise the enemy, attack him from unexpected directions, and destroy him with a violent, precise attack. The maze of cellars, rooms, sewer systems, rooftops, and alleys common to urban areas presents many opportunities for the ranger force.

b. Offensive operations by a ranger force exploit the centralized planning and decentralized execution aspects of ranger unit training. Individual action and initiative are always channeled to accomplish the commander's intent. The ranger unit's offensive tactics are based on quick action and boldness. Attacks by the ranger force are short and sharp. Execution is marked by swiftness, daring, timing, and surprise. The enemy's sleeping habits, eating habits, and relief times are watched. launched almost always at night or under smoke screens, attacks are often made without mortars or artillery, relying on surprise for success. Snipers and sharpshooters are used to shift the balance of battle.

c. Deliberate defensive operations are not good ranger unit missions. However, if a ranger unit must defend for a short period, the close terrain of urban areas allows it to function well against a stronger force. The buildings in a city break up advancing enemy formations and force them to move along streets. The ranger force executes an active defense anchored on strongpoints, such as a building or a group of buildings. It launches surprise counterattacks against the enemy's flank and rear. The aim of ranger defense is to give the enemy no respite and to make him feel threatened by a sudden, deadly attack. Ranger units should be withdrawn from defensive positions in urbanized areas as soon as possible. Prolonged MOUT and the associated casualties quickly degrade a ranger unit's combat effectiveness.

6-5. Perimeter defense.

The Perimeter defense may be conducted to control key terrain, gain time or economize forces.

a. **Planning.**

(1) **Warning order.** The order must be issued soon after mission receipt. It should contain the mission, time schedule (including rehearsal and inspection times, mission time, and place for OPORD), uniform and equipment, and special instructions.

(2) **Operation order.** The leader normally gives the OPORD on the ground to be defended. Locations and sectors of fire for squads and crew-served weapons should be designated. If time permits, leaders walk the terrain with subordinates as the OPORD is issued. The OPORD covers security, communications, signals, target reference points (TRPs), armor kill zones, and priority of engagement rules for antiarmor weapons. The OPORD also states the priority of work.

(a) Set up security and communications.

(b) Position crew-served weapons and personnel.

(c) Clear fields of fire.

(d) Prepare fighting positions.

b. **Organizing.**

(1) Squads are positioned with overlapping and interlocking sectors of fire and observation.

(2) The M60 machine guns are positioned either with an assigned final protective line (FPL) of fire or with a primary direction of fire (PDF). The PI)Fs are employed on covered avenues of approach such as gullies. The M60 FPLs should interlock across the unit front.

(3) The 90-mm recoilless rifles and Dragons are positioned with frontal cover to engage attacking armored vehicles from the flank.

(4) Riflemen are positioned to support and protect crew-served weapons. Riflemen are assigned FPLs that interlock at a point just beyond hand-grenade range.

(5) Dead space is covered with Claymore mines, antipersonnel mines, and M203 fires.

(6) Antitank mines, positioned to provide close-in protection from armored vehicles, are covered by LAW fires.

(7) Leaders plan indirect fire for long-range suppression of armored vehicles and short-range protection from dismounted assault. Mortar or artillery final protective fires (FPF) are placed on the most dangerous dismounted avenue of approach.

(8) Range cards are prepared for all crew-served weapons. Squad and platoon leaders prepare sector sketches and fire plans.

(9) The ranger force camouflages everything from ground and aerial observation.

(10) Leaders designate alternate and supplementary positions.

c. **Executing.** Leaders control the distribution and concentration of fires in their sectors. They constantly consider the overlap of fires with units to their left and right.

(1) Squads and crew-served weapons engage targets in sectors assigned by platoon leaders.

(2) Each weapon is used in its best role:

(a) Rifle, against enemy infantry within 300 meters.

(b) Machine gun, against enemy infantry within 1,000 meters.

(c) Grenade launcher, against infantry within 350 meters and armored vehicles within 200 meters.

(d) Light antitank weapon, against armored vehicles within 200 meters.

(e) Recoilless rifle, against armored vehicles within 400 meters.

(f) Dragon, against armored vehicles within 1,000 meters.

(3) Most dangerous targets are engaged first.

(4) Squads shift fires as needed or as directed by platoon leaders.

(5) The platoon masses fire on selected targets while maintaining coverage of the entire platoon sector.

(6) Platoon leaders control fires.

(7) After repelling an enemy assault, the platoon consolidates and reorganizes the defense.

6-6. Breakout from encirclement.

A breakout from encirclement is conducted when units, operating behind enemy lines, find themselves cut off from friendly forces and surrounded by superior forces. Given ranger unit missions, the chance of operating as an isolated force behind enemy lines is great. The breakout is characterized by determining enemy weak points, deception, massing of combat power, and a direct attack for a violent and timely breakout.

a. **Characteristics and fundamentals.** The breakout from encirclement involves the following characteristics and fundamentals:

(1) Set up a hasty defense. Adjust the perimeter as needed and make contact between subunits. Assign each a specific sector.

(2) Plan fire support. Plan FPFs on likely dismounted avenues of approach. Mass antiarmor weapons on the most dangerous mounted avenue of approach.

(3) Conduct reconnaissance. Identify enemy weak points. Locate enemy armored forces and the routes they must use. Locate enemy automatic weapons.

b. **Planning.**

(1) Organize all support personnel into maneuver elements--for example, company headquarters, mortars, and antitank. Organize the ranger force into a rupture force, reserve, support force, and a rear guard.

(2) Issue a fragmentary order. It should contain current local enemy situation and cover all pertinent areas of the five-paragraph field order.

(3) Select rally points.

c. **Execution.**

(1) Reduce the defense perimeter. This is done slowly until the unit is massed for the breakout attempt.

(2) Create a diversion. This is done either by fire from the rear guard force or by a limited attack in a direction other than that of the main breakout attack.

(3) Break through the defense with the rupture force. Hold the shoulders of the breakthrough.

(4) Move the reserve through to exploit the breakthrough.

(5) Move the support troops through with casualties. Have them collect casualties en route.

(6) Have the rear guard follow the support troops. It may have to fight a delaying action and use fire and movement to break contact with the enemy.

(7) Consolidate and reorganize following the breakout.

(8) Have the rupture force break contact with the enemy once the rear guard joins.

6-7. Linkup operations.

During operations behind enemy lines, ranger units may be required to link up with other friendly ground forces. Linkup may occur in airborne, air assault, amphibious, or shore-to-shore operations. It may occur during the breakout of an encircled force or during the convergence of separate forces. It may involve conventional or unconventional US forces, allied forces, or friendly guerrillas. Rangers may be required to link up with other rangers, partisans, or friendly ground forces that are conducting offensive operations.

a. **Planning.** Ranger commanders plan detailed linkup operations to avoid friendly forces firing on each other. Coordination and planning of communications, recognition signals, fires, command relationships and responsibilities, and control measures are essential. Coordination should be done before the start of the operation.

(1) Site. The linkup site should be easy to recognize, have cover and concealment, be located away from prominent terrain the enemy might use, be defensible, and provide multiple access and escape routes. An alternate site should be planned in case of compromise.

(2) **Communications and recognition signals.** Frequencies, call signs, codes, visual signals, and alternates for each of these should be planned before departure of friendly lines. Radios may be used to report the location of each unit to the site, and occupation and securing of the site. A system of mutual recognition must be devised to keep friendly patrols from firing on each other. This may include visual signals such as arm bands, panels, colored lights, or RSTA devices.

(3) **Fires.** The headquarters directing the linkup sets up fire control measures and priority of fires. The most commonly used control measure is the restrictive fire line (RFL), beyond which one force may not fire without coordination with another force. The RFL may be adjusted as two units draw closer together. Such an adjustment needs close coordination and should be planned before the operation. Successive phase lines between the two units can serve as on-order RFLS, provided that neither unit comes too close to the RFL before it is shifted.

(4) **Command relationships and responsibilities.** These are set up by the headquarters directing the linkup. Ideally, liaison personnel are exchanged before the operation. The key to success is liaison and guides.

b. **Execution.** Each member of the ranger force must understand the plan. Rehearsals are necessary in order to avoid confusion during the linkup. One uninformed ranger can start firing and cause enough confusion that friendly forces could be identified as enemy. Actions must be quick since units are open to enemy attack while they try to link up.

(1) **Moving units.** A linkup between moving units is hard. As the units move closer to each other, the chance of them engaging one another increases. Therefore, the linkup units must adjust their movements to each other and continuously coordinate on a designated secure radio net. If possible, one or both units should come to a short halt before linkup.

(2) **Stationary units.** Moving units must know the positions of the stationary units. They must keep stationary units advised of locations. Stationary units guide the moving units to the linkup point by radio. The stationary units must be ready to accept and position the moving units smoothly and quickly.

6-8. Passage of lines.

A passage of lines is an operation in which one unit moves through another unit that is stationary and disposed in a tactical formation on a FEBA. Ranger forces may conduct a passage of lines to get behind the enemy. Movement in forward unit areas must be controlled, coordinated, and kept to a minimum. This avoids conflict with friendly troops or the activation of RSTA devices. Ranger units treat the positions of forward units as danger areas. They must be assumed to be under enemy surveillance in all weather or visibility. The ranger force is open to attack during the passage. Personnel and units may be more concentrated than is desired. Fires of the stationary unit may be masked for a limited time. The passing ranger unit may not be in a good position to react to enemy contact. Detailed reconnaissance and coordination are crucial to ensure that the passage is conducted quickly and smoothly.

a. **Principles.**

(1) **Coordination.** The ranger force commander, or his representative, must coordinate departure and reentry of friendly front lines (FFL). This is done either directly with the units through which the rangers pass or through designated staff agencies. Items to be coordinated include fire support, passage lanes, control measures, guides, communications, intelligence, CSS, and actions on enemy contact.

(2) **Reconnaissance.** The ranger force commander should make a ground reconnaissance of the area through which the unit will pass. He should pick the initial rally point (IRP) at this time. The reconnaissance should include observing the area just forward of the friendly unit for routes, danger areas, and obstacles. During reconnaissance or coordination, rangers should wear the same uniform as the troops in the forward unit. This will avoid attracting the enemy's attention.

(3) **Security.** Maintain security to avoid contact with the enemy while departing or reentering. It is important that the ranger force make no enemy contact at this time. Control is hard if a fire fight starts just forward of an FFL.

b. **Departure of friendly front lines.**

(1) Designate an IRP.

(2) Maintain security. Use an appropriate movement formation when departing the FFL.

(3) Do not move within the friendly unit's area without a guide.

(4) Make final coordination with the friendly unit commander.

(5) Count the ranger force members as they depart.

(6) Make a security and listening halt.

c. **Departure of friendly units.**

(1) The ranger force arrives at the forward unit and is met by a guide. The guide leads the rangers to an assembly area chosen by the ranger commander during initial coordination with the forward unit commander.

(2) No one should move, either singly or as a unit, anywhere in the FFL without a guide.

(3) The ranger commander makes final coordination with the forward unit commander. He determines if any changes have taken place since initial coordination and learns of any recent enemy activity.

(4) Before leaving to coordinate, the ranger commander gives instructions (called a contingency plan) for what should be done while he is gone. These instructions state where he is going, who is going with him, how long he will be gone, what to do if he does not return, and actions to be taken if there is enemy contact.

(5) On the commander's return from final coordination, he may issue a fragmentary order (FRAGO) to cover any changes.

(6) Three common threats and techniques for countering them during departure from the FFL are:

(a) **Ambush and chance contact.**

- The ranger force commander must avoid enemy contact while departing.
- The ranger force goes to the friendly side of the departure point where a point team is dispatched to see that the area forward of the barriers is clear.
- The point element should check the area out to the first covered position or an area large enough to allow the unit room to maneuver if it is hit while departing.
- The distance varies with the size of the ranger force and the terrain.
- The point team notifies the ranger commander when the area is clear. The rest of the unit is brought forward.

(b) **Indirect fire.** If enemy indirect fire is falling, the ranger commander should not halt his unit after final coordination at the forward command post. Instead, he should move it quickly through the FFL. The point element should be far enough ahead of the main body to provide security from ambush or chance contact.

(c) **RSTA.** If the enemy has radar, sensors, or night vision devices, they may be offset by the following countermeasures:

- Use a well-hidden departure point, such as a reverse slope or a dense woods.

- Infiltrate the unit through the departure area and have the rangers rendezvous in a defilade.
- Pass through during rain, fog, snow, or darkness to help conceal the passage.
- Employ electronic countermeasures.

(7) Have a security and listening halt after the ranger force has moved out of sight and sound of the FFL. This is a short halt to see if enemy is in the area, and to accustom the unit to the sights and sounds of the battlefield. Halt in a position that covers the unit from chance friendly small-arms fire. If the security and listening halt must be in an exposed position, have each man lie flat on the ground.

(8) During movement, have frequent halts to see that the unit is not being followed and that no enemy is in the area.

d. **Reentry of friendly front lines.**

(1) Set up and occupy a reentry rally point (RRP).

(2) Maintain security at the RRP and at the reentry point.

(3) Use as few personnel as possible to locate the reentry point while the bulk of the unit remains in the RRP.

(4) Meet a guide at the reentry point. Normally, a forward-of FFL password is used since it may be overheard by enemy.

(5) Count the ranger force members to ensure no infiltrators follow the unit, especially during reduced visibility.

(6) Give the friendly unit commander a spot report, providing combat information of immediate tactical value.

e. **Reentry of friendly units.**

(1) The unit is moved into a rally point near the reentry point. This rally point should be on a prominent terrain feature where the ranger commander can pinpoint his location, with respect to the reentry point.

(2) The forward unit is alerted by radio that the ranger unit is ready to reenter. A code word is used for security and brevity. The code word must be acknowledged by the forward unit before the ranger force reconnoiters the reentry point. This shows that a guide has been sent to the reentry point and is waiting for the unit.

(3) If the ranger commander is certain of the reentry point location, he moves the entire unit directly to the reentry point. He has the point element coordinate the reentry.

(4) An effective method of locating the reentry point is through the use of a thermal imagery device. It detects the body beat of the reentry guide, even though he may be well camouflaged or his location masked by smoke. If ground surveillance radar is available, it can be used to vector the ranger unit to the reentry point.

(5) The reconnaissance party should never move parallel to friendly barriers or move around the wire. If the reentry point cannot be found by the initial reconnaissance or if the reconnaissance party finds only the barrier wire, the unit should notify higher headquarters and move to another rally point to wait until daylight. The unit should not stay in a rally point from which a radio transmission has been made.

(6) When the reentry point is found, the ranger commander can bring the unit forward. If the reentry point is easy to find, he can call by radio for the unit to come forward.

(7) The guide leads the unit through the barriers to the security position coordinated with the forward unit commander. The unit halts in the security position, remaining in movement formation. The ranger commander then gives a spot report to the forward unit commander. The ranger commander tells him anything of intelligence value or of immediate tactical use to the forward unit.

6-9. Reconnaissance platoon operations.

a. The reconnaissance platoon organic to the ranger regiment consists of a platoon leader, a senior reconnaissance sergeant, and three five-man reconnaissance teams that include a communications specialist in each team (see Appendix A). This platoon is responsible for short-duration reconnaissance of objectives for the ranger force commander. This reconnaissance is routinely in support of future strike or special light infantry operations. The teams can also be used by the ranger force commander to conduct deep target acquisition and designation, or to confirm or disprove information reported by other sources. Rarely will teams be used to attack enemy targets.

b. The reconnaissance teams of the ranger regiment do not replace (or perform the same reconnaissance tasks associated with) long-range reconnaissance patrol (LRRP) units or long-range surveillance units (LRSUs) that may be organic to the corps or division. These units are organized, trained, and equipped for long-term passive surveillance and reconnaissance in enemy territory. The ranger reconnaissance teams are trained and organized to support the immediate intelligence-gathering needs of the ranger force commander. They are not to act as a human intelligence (HUMINT) collection asset for other operational headquarters.

c. The reconnaissance platoon headquarters trains, equips, and controls the employment of the three reconnaissance teams. It operates the communications stations needed to receive reports from deployed teams. In the absence of the S2, it debriefs team members upon extraction or exfiltration. It can provide one or more teams to support each ranger battalion on independent operations. The reconnaissance teams can--

(1) Infiltrate the objective area by parachute (HALO, HAHO, or static lines), helicopter, fixed-wing aircraft, SCUBA, small boat, foot, or other means.

(2) Remain undetected in the vicinity of the objective area up to five days.

(3) Perform reconnaissance operations employing a full range of night observation devices, infrared detection devices, unattended sensors, and photographic equipment.

(4) Perform demolition target analysis.

(5) Operate small watercraft and inflatable boats.

(6) Emplace unattended ground sensors, omni-directional navigational beacons, hand-emplaced expendable jammers, and electronic target designation devices.

(7) Collect combat information to satisfy priority information requirements and mission-essential elements of information. Teams report that information by use of long-range, secure, burst-transmission communications equipment.

(8) Perform DZ selection, marking and reception duties.

(9) Report objective area weather conditions.

(10) Perform highly selective, limited attacks or ambushes when so tasked.

(11) Link up with the main body of the ranger force in the objective area, or escape and evade the enemy in order to return to friendly lines.

(12) Act as part of the AAT during airborne operations.

d. There are many limits of the reconnaissance teams.

(1) Mobility is normally restricted to foot movement in the area of operations.

(2) Use of radio and other active electronic or optical devices makes the teams open to enemy detection.

(3) All supplies and equipment needed by the team must be carried on the initial insertion, since resupply increases the probability of detection by the enemy.

(4) Conduct of insertion and extraction operations deep in enemy-held territory is difficult due to the need for secrecy, timeliness, security, and accuracy of location.

(5) Medical treatment of team casualties is limited to individual first aid. Casualty evacuation is difficult unless linkup with the main ranger force has been achieved.

e. The training of reconnaissance team members parallels that given to all members of ranger units with emphasis on infiltration, communications, and reconnaissance skills. Extensive training is also given in survival, evasion, resistance, and escape (SERE) and advanced first aid. Training emphasis is placed on actions at the reconnaissance site. The reconnaissance teams must conduct their mission and remain unseen. Every team member is trained to consider the need for information balanced by the likelihood of detection. Detection by the enemy compromises future operations.

6-10. Reconnaissance planning.

a. Detailed planning ensures that the reconnaissance mission is successful. It is also vital to the survival of the reconnaissance team. This planning is conducted by the ranger force planning group assisted by elements of the JTF staff. The supporting air or naval commander, navigator, and electronic warfare officer may aid the ranger force staff in the planning.

b. Upon receipt of the mission, the team selected begins an intense preparatory phase at the REMAB or the ISB. There the team receives its initial mission briefing. The items covered allow the team to complete its detailed planning. The reconnaissance platoon leader reviews the team plan. He ensures compatibility and integration into the ranger force commander's plan for employment of the main force. Staff specialists at ranger regiment and battalion--such as those concerned with intelligence, weather, or communications--provide help.

c. Planning for the use of reconnaissance teams is conducted within the framework of the ranger planning sequence for the entire mission. There are five planning considerations for the employment of reconnaissance teams:

(1) Operations security must be considered in all staff efforts to provide maximum protection for the reconnaissance teams. This includes intelligence, CE, logistics, administration, and maintenance. Operations security must be integrated throughout every ranger force operation. It consists of four main categories of security measures: signal security, physical security, information security, and deception. All are related and must be considered at the same time for each operation.

(2) Some missions may require rapid deployment to the operational area, thereby dictating the quickest method of insertion. However, others may depend on maintaining total secrecy, regardless of time.

(3) The transportation means selected for the delivery and recovery of the teams depends on the mission. This selection is also based on the abilities, limits, and availability of mission support platforms.

(4) The distance to and from the objective area must be considered, as well as the distance from the insertion point to the actual target. Reconnaissance teams can move great distances over rugged terrain. However, this movement does require time and limits the team's ability to carry bulky or sensitive equipment.

(5) Consideration must also be given to intelligence information available concerning the objective area which includes, but is not limited to--

(a) **Enemy situation.** The enemy threat--his abilities, disposition, security measures, detection, and air defense systems - affects the selection of the means for delivery or recovery.

(b) **Terrain.** Land formations must be considered in selecting the method of infiltration and exfiltration. Terrain affects the selection of altitudes, approach and exit routes, landing areas, drop zones, and beach landing sites. Air infiltration routes that provide terrain masking are desirable in static line parachute operations.

(c) **Weather.** Seasonal weather conditions affect team delivery and recovery. Factors to be considered are temperature, precipitation, visibility, clouds, and wind. If para-SCUBA techniques are used, high surface winds and their effect upon surf conditions or period of reduced visibility may preclude the use of parachutes, inflatable boats, or swimming as entry or recovery techniques. These same conditions favor land infiltration and exfiltration, The AWADS reduces the impact of visibility as a limiting factor for air infiltrations.

(d) **Astronomical conditions.** Periods of sunrise and sunset, moon phase, moonrise and moonset, and twilight must be considered.

d. A reconnaissance team mission must be specific and support the mission. Information concerning the target may not be gathered because the risk of detection by the enemy is so great. Detection may preclude the conduct of the planned operation. If more than one mission is assigned, priorities are set, All team missions must be coordinated to avoid other friendly forces operating near the target. Ranger forces conduct operations at such depth behind enemy lines that no other US or allied forces are normally present. There is a chance that a friendly indigenous guerrilla force may be operating near the objective, but ranger reconnaissance teams do not normally make contact with them.

e. The reconnaissance platoon leader joins with the intelligence and operations sections of the ranger force headquarters in the initial planning for reconnaissance team operations. The method of operation while moving, communication procedures, reporting, surveillance procedures, and actions on enemy contact should be included in the platoon's SOPS.

f. The ranger force intelligence officer normally prepares the reconnaissance plan IAW the guidance provided by the controlling headquarters. He is helped by the operations officer and the reconnaissance platoon leader. Selected team members and a member of the unit providing the insertion and extraction means are briefed on the mission early in the planning phase. They should participate in the detailed planning to follow. The essential details of a reconnaissance plan normally include--

(1) Areas to be observed and recommended areas or positions from which this can be done. These positions are determined in advance of the insertions of teams. They are based on a study of terrain, road and rail nets, enemy order of battle, delivery means, operations plans of the controlling headquarters, and the desires of the ranger force commander.

(2) Load plans and procedures. This includes both insertion and extraction. Alternate sites are chosen for possible use. Primary and alternate DZs are selected, if applicable.

(3) The flight plan. This includes insertion and extraction flight routes and alternate routes.

(4) The movement plan to and from a patrol base. This includes the times, routes,. and destinations when individuals or small groups are required to operate away from the patrol base.

(5) The fire support plan. This includes the location and times of planned air strikes in support of the main body, and all planned fires by the main assault force. The reconnaissance teams should know the frequencies and call signs for tactical air support.

(6) The timing and ground tactical plan for execution of the main force operation. This includes disposition of the main force, special mission teams from the ranger force, or any other special operations force elements.

(7) The communications plan. This includes frequencies, reporting schedule, emergency reporting procedures, and alternate communications plans. It also includes the ground-to-air

communications plan and the friendly recognition signal to be used during linkup with the main assault force or other friendly forces.

(8) Plans for use of technical specialists or special equipment.

(9) The plan for the treatment and evacuation of sick or wounded team members from the objective area.

(10) The E&E plan. This includes routes out of the objective area, passwords, and alternate extraction zones or methods far from the objective area.

(11) Mission-specific elements of information desired by the ranger force commander, and any special or on-order missions the team must conduct.

g. The team leader uses specific steps in planning, preparing, and executing missions. These procedures are comprehensive, yet adaptable to any situation. Three of the most important steps are:

(1) **Issuance of warning order.** This action gives the team maximum time to prepare for the mission. It should be issued as soon as the team leader has been notified of a mission and updated after he has made his tentative plan.

(2) **Issuance of OPORD.** The team leader conveys the information and instructions needed to accomplish the mission. He requires a briefback from each team member.

(3) **Rehearsals.**

h. The specific steps used when planning missions are as follows:

Step 1. Study the mission.

Step 2. Plan use of time.

Step 3. Issue initial warning order.

Step 4. Study terrain and situation.

Step 5. Make tentative plan.

Step 6. Organize.

Step 7. Update warning order.

Step 8. Coordinate (continuous throughout).

Step 9. Consult aerial photographs and map updates.

Step 10. Complete detailed plan. (To be combined with paragraphs 3, 4, and 5 of the operation order.)

Step 11. Issue operation order.

Step 12. Supervise (throughout), inspect, rehearse.

Step 13. Execute mission.

i. A workable E&E plan must be developed. Such a plan enhances the survival of the reconnaissance team members. It ensures that they are psychologically prepared to conduct the mission, by knowing there is a system and plan for returning them to friendly lines. The E&E plan must include a section describing the actions of the reconnaissance team members should their insertion craft be disabled en route to the objective area. It should also include what to do if the team is unable to link up with the main ranger force, or if the main ranger force has not been inserted into the objective area.

j. Rehearsals and briefbacks are best for finding flaws in procedures or planning. Thorough coordination of all procedures is essential. The exact type of aircraft or other means that will be used for insertion should be used for training. Rehearsals should be conducted under terrain, astronomical, hydrographical, and meteorological conditions close to those to be met on the operation. The team leader conducts a briefback with the platoon leader. This lets the platoon leader decide if the team is ready for the mission. It also allows him to make changes.

6-11. Execution of insertion.

a. Reconnaissance teams normally conduct blind insertions. Blind insertions are amphibious, heliborne, or parachute drops onto unmarked DZs without help from a DZ-marking party or a CCT. Blind insertions often depend on favorable light and weather conditions in the objective area. Favorable conditions for the reconnaissance team often means low-lying cloud cover, fog, and darkness. The teams are able to operate during reduced visibility by the extensive use of night observation devices. The team's extensive training and high level of land navigation skills allow it to quickly traverse rugged terrain and not be detected.

b. Team insertion into the objective area often needs a cover and deception plan. Team members can conduct HALO and HAHO operations. Teams can be inserted into areas not normally acceptable as a DZ for conventionally trained parachutists. By the use of HAHO techniques, reconnaissance teams can assemble in the air and glide to a landing at a location far from the drop point.

c. If reconnaissance teams need special or bulky equipment that cannot be carried on the individual team member, it can be airdropped, hidden near the DZ, and retrieved later. Such container drops are often used during low-level static line insertions. The high-altitude airdrop resupply system (HAARS) can be used to deliver such items at the same time as reconnaissance teams are inserted by HALO.

CHAPTER 7

Combat Support

Combat support is used to enhance the combat power of the maneuver companies. Combat support elements may be organic or nonorganic. They can conduct a variety of missions in providing indirect and direct fire support, security, intelligence, and maneuver and fire control.

7-1. Intelligence support.

a. The success or failure of a ranger mission often depends on the accuracy, detail, and timeliness of intelligence information. Both deliberate and quick-response missions need specific information that can be obtained only by a multidiscipline collection and analysis agency. An active interface with all levels of the intelligence system helps get the needed information quickly.

b. The theater or JTF commander is responsible for providing intelligence support. The 1st SOCOM also provides information and analyzed intelligence to the ranger force commander. The main source of processed intelligence is the Echelons Above Corps Intelligence Center (EACIC), similar to the division level military intelligence (MI) all-source production center (ASPC). The EACIC provides intelligence, security, and electronic warfare (EW) interfaces (see Figure 7-1). It is through the EACIC that ranger commanders can request support from national, joint, combined, or CONUS intelligence analysis systems.

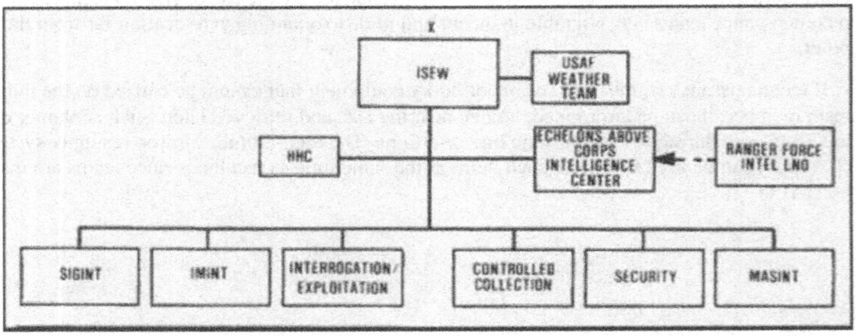

Figure 7-1. Intelligence at echelons above corps.

c. Intelligence requirements for ranger operations are unique, highly sensitive, and compartmented. The intelligence officer of the 1st SOCOM and the appropriate 1st SOCOM regional planning cell coordinate with theater intelligence agencies to provide needed information.

d. The ranger regimental staff and the ranger battalion staff have personnel in their intelligence sections to correlate and analyze information from many sources.

e. The reconnaissance platoon, organic to the ranger regiment, maintains surveillance of the objective area before the insertion of the main ranger force. It reports any last-minute developments that would affect the ground tactical plan. The platoon can also be tasked to determine any mission-specific items of essential information that the ranger force commander needs to complete his plan. This platoon is a highly specialized unit that is reserved for use by the ranger force commander. It is not a HUMINT collection agency for use by the theater or JTF commander. Reconnaissance teams normally link up with the ranger force in the objective area and are extracted with the main body. If this linkup cannot be made, they will exfiltrate enemy territory. They may use escape and evasion nets set up by other special operations forces or Department of Defense (DOD).

f. The ranger regiment normally places a liaison team at the headquarters of the command having OPCOM of a ranger force. This team provides command and control communications and liaison. It puts an intelligence liaison officer (LNO) at the EACIC or the appropriate ASPC (see Figure 7-2). This provides coordination to ensure that the intelligence needs of the ranger force are relayed to the intelligence processing center, and that the analysis is on the specific needs of the ranger force commander. This LNO coordinates with another located at the corps or EAC targeting center. Close coordination between the regiment, the intelligence analysis center, and the targeting center is essential to effective employment.

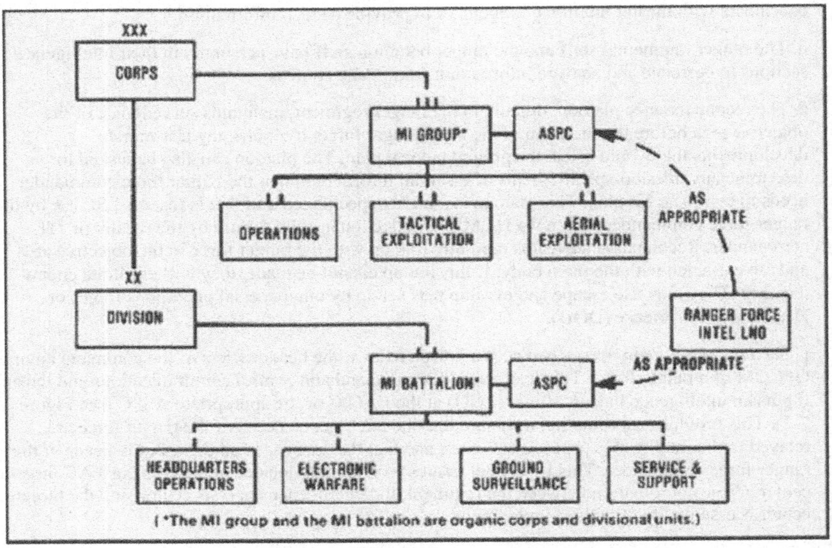

Figure 7-2. Intelligence source at corps and division.

7-2. Electronic warfare support.

a. Depending on the nature of the target and enemy abilities, a ranger mission may need support from EW units. This is mainly true during insertion. Electronic warfare operations disrupt or destroy the enemy's command and control of his forces and weapons systems, and retain friendly use of the electromagnetic spectrum. Electronic warfare is also used to support deception operations. They mislead the enemy by manipulation, distortion, or falsification of indicators, and persuade him to react in a manner prejudicial to his own interests.

b. Active jamming and chaff dispersal can prevent enemy early-warning radar from detecting the airborne force or determining its route. False transmissions can aid deception plans or feints used to help the ranger force's entry behind enemy lines. Selective jamming and imitative deception can disrupt the enemy command and control nets. This prevents the enemy from reacting to the presence of the ranger force in time to prevent mission accomplishment, or slows and interrupts his deployment of reaction forces.

c. Ranger elements, or other military forces, may conduct closely coordinated direct attacks against enemy EW sites during a ranger force's insertion or extraction. Not only does this reduce the enemy's ability to direct EW against the ranger force command, control, and communication system, but it misleads and confuses the enemy as to the true ranger objective.

d. The JTF or theater commander is responsible for planning the EW program to support a ranger mission. Planning and coordinating elements from the 1st SOCOM may also help. The EW program must be closely coordinated so as not to disrupt friendly air-to-ground or ground-to-ground communications, or to reveal the ranger force intentions.

e. The ranger regimental communications-electronics (CE) officer and those of the ranger battalions ensure that the required coordination concerning jammed frequencies, codes, authentication tables, and so on, is completed. The ranger regimental and battalion operations personnel ensure that the efforts of all EW units are combined to support the ground tactical plan.

7-3. Communications support.

a. Ranger operations are supported by secure long-range, lightweight, real-time, high-frequency (HF), and satellite communications. Effective long-range communications provide command and control links between deployed ranger units and the controlling headquarters. The JTF, ARFOR, or theater commander is responsible for communications between the controlling headquarters and the ranger regiment. The special operations force commander may also provide secure communications terminals to the ranger regiment or a deployed battalion.

b. Secure AM, FM, and SATCOM radios are the primary means of communication within the ranger regiment. Within the ranger battalion, AM and FM radios provide communications to company, platoon, and squad level.

c. During certain missions, specially trained and equipped quickreaction elements (QRE) deploy with the ranger force to provide secure communication links to the special operations force commander. These JCSE elements operate on either SATCOM or tactical satellite terminal (TACSAT) channels or through an airborne communications relay platform (see Figure 7-3). Depending on the mission, an airborne battlefield command and control center (ABCCC) aircraft or a joint airborne communications center/command post (JACC/CP) may be used. Their communications systems can operate at all levels of the national chain of command to permit a quick response to the tasking authority.

Figure 7-3. Airborne communications relay platform.

d. When only one ranger battalion is employed and the regimental headquarters is not the controlling headquarters, the ranger regiment provides a liaison cell to the controlling higher headquarters. This liaison cell includes a communications element from the regimental communications platoon. It can provide secure SATCOM, teletype, and facsimile transmission support to the ranger force in the objective area.

e. If two or more ranger battalions are employed, the ranger regimental headquarters deploys and acts as the command and control headquarters. The regimental communications platoon would then provide another communications link to the special operations force commander.

f. The ranger regimental CE officer ensures that the necessary communications links are set up and coordinated. The many communication means and channels available provide for effective control of a deployed ranger force. However, communication means must be closely coordinated at all levels of command to control the complex operations of a ranger mission. Planning and coordination with supporting aviation, transport, fire support, medical, and logistical elements before an operation are vital to efficient communications. The regimental CE officer must consider the communication systems linking the ranger force and other services. Air-to-ground and ship-to-shore communications are vital and must be set up early in ranger operations.

7-4. Fire support.

a. **Planning.**

(1) The ranger regiment has limited organic fire support assets. The six 60-mm mortars in each ranger battalion (two in each rifle company) provide the only organic indirect fire

available to the battalion commander. Each rifle company also has three 90-mm recoilless rifles and three Dragon missile systems that provide a small measure of organic direct fire support.

(2) The missions assigned to the ranger regiment and its subelements normally require fire support from organizations outside of the regiment. To plan, coordinate, and control these fires, the ranger regiment is organized with personnel dedicated to these tasks.

(3) The ranger regimental headquarters has a fire support element consisting of one fire support officer (FSO), two noncommissioned officers, and two fire support specialists. They prepare the fire support annex to the regiment's operation order. They ensure that all available fire support is planned to best support the regimental commander's ground tactical plan. They act as the primary coordinator for all external fire support, such as artillery, armed helicopter, close air support, naval gunfire, and short-range air defense. They also coordinate with each ranger battalion FSO to ensure unity of effort. The fire support element coordinates the efforts of air force, naval, and marine liaison officers when they are attached.

(4) The regimental commander may assign part of the fire support element to a committed ranger battalion as an attachment to help in the integration and planning of fires. The ranger battalions request through the fire support element that more or different fire support means be integrated into the ground tactical plan.

(5) The ranger battalion headquarters has a fire support team headquarters consisting of the battalion's FSO, two noncommissioned officers, and two fire support specialists. The FSO is the prime fire support coordinator for the ranger battalion commander. He maintains coordination with the regimental FSO and any attached fire support coordination elements, such as the USAF tactical air control party or naval shore fire control party. The FSO performs the following:

(a) Advises the ranger battalion commander on all fire support matters.

(b) Recommends allocation of fire support.

(c) Prepares fire support plans.

(d) Assigns target numbers.

(e) Processes target lists from the ranger company fire support teams to eliminate duplication.

(f) Monitors and functions as net control on the ranger battalion fire support coordination (FSCOORD) net.

(g) Reports changes in the status of fire support units to the ranger battalion commander, staff, and FISTS.

(6) When the ranger battalion is operating a single TOC, the FSO and his element are normally located there. When the ranger battalion is operating two TOCS, one TOC is normally manned by the FSO, the fire support sergeant, and a fire support specialist. The

senior fire support sergeant and the other fire support specialist would then monitor the FSCOORD net at the other TOC. The FSO ensures that any attached fire support elements also divide when needed to operate in a two-TOC configuration.

(7) The ranger battalion fire support team headquarters has three fire support teams assigned to it. These are normally allocated one to each of the ranger rifle companies. The company FSO and his team--

(a) Locate targets and request and adjust surface-to-surface fire support (mortar, field artillery, and naval gunfire).

(b) Plan fires to support the company ground tactical plan and prepare target lists.

(c) Coordinate fire support requests through the FSO for surface-to-surface and air-to-surface fires.

(d) Report battlefield information.

(e) Control air-to-surface fires of helicopters and prepare to assume control of fires from fixed-wing aircraft.

(f) Coordinate airspace use within the ranger rifle company operational area.

(g) Advise the ranger company commander on all fire support capabilities, limitations, and methods of employment.

(h) Inform all fire support units of target priority changes.

(i) Designate targets for laser-guided munitions.

(j) Assume operational control of, and administrative responsibility for, all augmentation fire support personnel (TACP, shore fire control party, and attack helicopter liaison teams).

(8) The company FSO is normally with the ranger rifle company commander. The FOs and their RATELOs are normally with the ranger rifle platoon leaders.

b. **Coordination.**

(1) Coordination of fires is a continuous process. It involves implementing the fire support plans of the ranger force commander and reacting to changing tactical situations. The success of a mission often hinges on how well fires are coordinated during the operation.

(2) The fire support planners within the many echelons of the ranger force become fire support coordinators during the actual execution of the operation. They use standard fire control coordination measures.

c. **Organic fire support.**

(1) **Indirect fires.**

(a) The 60-mortars constitute the only organic indirect fire support available within the ranger regiment. The mortar squads in each ranger rifle company normally operate as a section under the control of the mortar section sergeant. However, they can operate alone for short periods.

(b) The 60-mm mortar is used to destroy enemy equipment and light vehicles and to attack enemy personnel. Its main advantage is the ease with which it can be brought into operation against the enemy. This ease of employment allows the ranger force commander to concentrate combat power at the critical time and place. This helps to surprise and destroy the enemy.

(c) Mortar fires are employed by the ranger rifle company commander and coordinated by the company FSO. They are normally requested and adjusted by the FIST observers. The mortars are normally emplaced in a central position, and their fires are controlled by the fire direction center (FDC) using the indirect-fire method. The mortars may also use the direct-alignment or the direct-lay method.

(2) **Direct fires.**

(a) Each ranger battalion is authorized 90-mm recoilless rifles and Dragon missile launchers. These weapons constitute the heaviest organic direct-fire ability in the regiment. The antitank section in each ranger rifle company may operate as a complete section, but normally individual weapons are assigned throughout the company sector wherever their fires can best support the mission. These weapons can destroy most enemy armored vehicles and can penetrate the flank and rear armor of all enemy main battle tanks. The 90-mm recoilless rifle has a multiple projectile round that is effective against exposed enemy personnel. The HE round is highly effective against buildings and bunkers.

(b) The main disadvantage of direct-fire antiarmor weapons is their weight. These weapons, and their ammunition, are man-portable but place a strain on the unit when moving across rugged terrain.

(c) The ranger battalion is trained in the use of light antiarmor weapons, such as the US M72A2 LAW and the Warsaw Pact rocket-propelled grenade (RPG). The following provide protection for the ranger force against attack by enemy armored forces:

- The employment of light antiarmor weapons at all organizational levels.
- The placement of the 90-mm recoilless rifles or guided-missile systems on the most likely avenues of enemy-armor approach.
- An aggressive, offensively oriented employment.
- The ability to move into and across terrain impassable to armored vehicles.

d. **Field artillery fire support.**

(1) Many ranger unit operations are out of the range of supporting field artillery fires. Whenever such fires are available, they are planned for and integrated into the ranger force ground tactical plan.

(2) Coordination and exchange of call signs, frequencies, and target lists occur before insertion of the ranger force. Unless required by the tactical situation, ranger units do not normally receive augmentation forward observers from supporting field artillery units. The degree of mental and physical training necessary to execute certain ranger techniques requires more preparation than field artillery units are able to provide. The FIST from the ranger battalion requests and adjusts field artillery fires for the ranger force.

(3) Field artillery fires can be used to support the ranger force even if the objective area is out of range. Field artillery cannon fire and multiple rocket launch system (MRLS) fire can be used to suppress enemy air defenses. This helps the ranger force as it crosses the FEBA during airborne or air assault operations.

(4) Field artillery fires can be used to support the exflltration of ranger elements as they approach friendly lines. These fires can contribute to the deception plan and add combat power to feints used to support ranger operations.

(5) When in range of the objective area, field artillery units can be used to emplace field artillery delivered FASCAM to enhance the security of the ranger force.

(6) Copperhead rounds fired by 155-mm field artillery units can be terminally guided by the ranger FIST forward observers (FOs). They can attack hardened point targets or enemy armored vehicles by using a man-portable laser target designator (see Figure 7-4).

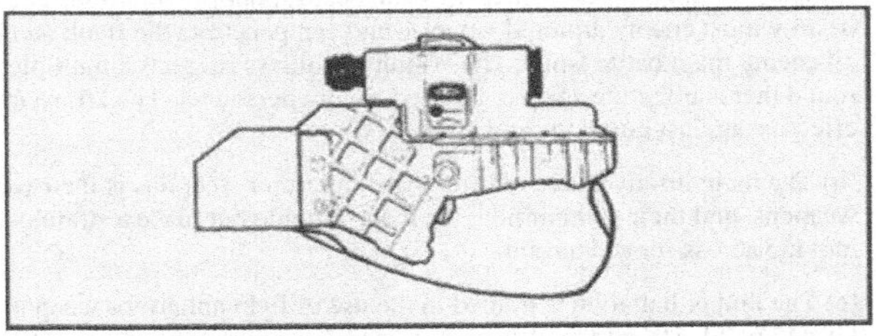

Figure 7-4. AN/PAQ-1 laser target designator.

e. **Aerial fires.**

(1) Aerial fire support is usually the prime means supporting the ranger force due to the distance behind enemy lines at which most ranger operations take place. Aerial fire support can be provided by either fixed-wing aircraft or helicopters.

(a) Fixed-wing aerial fire support may come from USAF, USN, or USMC units. The type of unit providing support, the aircraft, and the mix of ordnance carried all affect the fire support planning and coordination process. Some aircraft have a night and all-weather strike ability enabling them to support the ranger force during any level of visibility. Operations during poor weather that limits visibility to less than 3 nautical miles are still somewhat restricted. The ranger fire support coordinators must ensure that the correct aircraft are requested and employed effectively on the enemy. The

TACP directs and adjusts aerial fires in the objective area.

(b) The ranger force can use ground laser target designators to pinpoint targets for air strikes, as well as electronic navigation aids to permit nonvisual air strikes (beacon bombing). The ranger rifle company FIST or the TACP can control a laser-designated standoff air strike (see Figure 7-5).

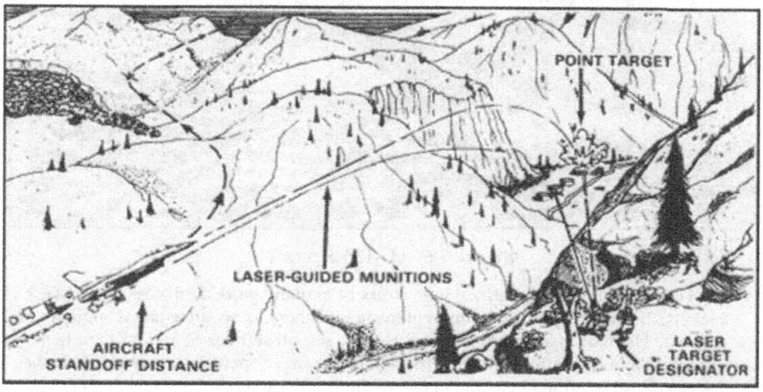

Figure 7-5. Standoff air strike.

(c) If the enemy ADA ability is not great or it can be degraded to a low level, the ranger force uses specially equipped and armed AC-130 aircraft (see Figure 7-6) for fire support. These aircraft provide an invaluable combination of firepower, night observation and illumination, communications, and long loiter time. A well-planned and executed suppression of enemy air defense (SEAD) program, coupled with ECM directed against enemy ADA units, normally permits the use of AC-130 aircraft.

Figure 7-6. AC-130 aircraft.

(2) The attack helicopter, armed with a mix of antitank guided missiles (ATGM), 2.75-inch rockets, 20-mm cannon, and 40mm grenade launchers, is accurate and responsive aerial fire support. The ranger force commander plans to use all sources of aerial fire to help accomplish his mission. However, the nature of ranger operations may preclude the extensive use of armed helicopters due to their limitations. Some attack helicopters are limited in range and lack an all-weather ability. They are restricted during operations at night by a lack of sophisticated night vision devices. They may lack the speed to go with the ranger force on some heliborne insertions. Attack helicopters may be used to escort and assist the ranger force as it crosses the FEBA. They may also be used to conduct feints and demonstrations to cover the insertion of the ranger force. As the MV-22 Osprey and the AH-64 Apache enter service, more use will be made of these aircraft to support ranger operations beyond the FERA, (See Figures 7-7 and 7-8).

Figure 7-7. MV-22 Osprey.

Figure 7-8. AH-64 Apache attack helicopter armed with Hellfire missiles.

(a) If attack helicopters are used to support a ranger operation, planned indirect fires are normally delivered along entry and exit corridors. Attack helicopters approach and depart the objective area using nap-of-the-earth flight profiles.

(b) Fires from armed helicopters are normally requested and controlled by the company FSO or one of his FOs, operating on a special ground-to-air net. The laser target designator may be used to precisely identify targets for the AH-64 Apache. Friendly unit locations may be marked by smoke, panels, lights, mirrors, or infrared sources.

f. **Naval gunfire.**

(1) During amphibious assaults or operations near a coastline, the ranger force may receive indirect fire support from naval gunfire. Normally, a ranger battalion would be supported by either a destroyer or a cruiser in a direct support role. However, the type and importance of the mission, the type of targets, the ships available, the hydrographic conditions, and the enemy capability determine how many and which type of ships are provided to support the ranger force.

(2) Naval gunfire is characterized by large volumes of highly destructive, flat-trajectory fire. Planned strikes in support of the ranger force may also include surface-to-surface missile fire. Some naval guns can fire a laser-guided projectile much like the Copperhead. This projectile, called a Deadeye round, is terminally guided the same as the Copperhead. Deadeyes can be guided either by an attached spotting party from the ANGLICO or by ranger LTD teams.

(3) When ranger units are being supported by naval gunfire, a shore fire control party is attached to the ranger force, This party normally consists of a liaison team and several spotter teams (forward observers). The liaison team is integrated into the operations of the fire support element (FSE) at the ranger battalion TOC. The spotter teams are attached to the ranger rifle companies. The shore fire control party LNO is the ship's representative to the ranger force commander, through the FSO. The spotter teams request and adjust fires from surface vessels. They can also request and control air strikes by carrier-based aircraft.

(4) The LNO and the spotter teams operate in the ground spot net, communicating with the ship by high frequency (HF) radio to request and adjust naval gunfire. The spotter teams communicate with the LNO using very high frequency (VHF) radios. The LNO also can communicate with aircraft using ultrahigh frequency (UHF) radios.

(5) Coordination and control measures that apply to naval gunfire are the same as for field artillery, with the addition of two terms, which are:

(a) Fire support area (FSA)--a sea area within which a ship may position or cruise while firing in support. It is labeled as FSA and numbered by a Roman numeral - for example, FSA VII.

(b) Fire support station (FSS)--a specified position at sea from which a ship must fire. This is very restrictive positioning guidance. It is labeled as FSS with a Roman numeral--for example, FSS VII.

g. **Air defense artillery.**

(1) Army air defense artillery units are rarely used to directly support the ranger force since

ranger operations normally take place deep behind enemy lines. However, it is possible that during the insertion or extraction phase of an operation, ADA units may support the ranger force as it crosses the FEBA.

(2) The ranger regiment's fire coordination element and the FSO in the ranger battalion are mainly concerned with adequate coordination of ADA fires. This prevents friendly units from being engaged. It allows the integration of ADA fires into the scheme of maneuver. The fire support planners ensure that enough coordination has been made so that ADA units are aware of the ranger force and know enough about its flight paths, routes, and altitudes to prevent engagement by friendly units. When conducting this coordination, care must be taken to pass only needed information to the ADA unit. To aid coordination, an ADA officer may be attached to the ranger force.

(3) The plan for a heliborne or airborne insertion and extraction of the ranger force must include the correct identification-friend-or-foe (IFF) codes, the flight profiles, the times of FEBA crossings, and the estimated number of aircraft making the crossing. There must also be a plan for crossing by straggler aircraft and battle-damaged aircraft that cannot reply to IFF interrogation. If possible, the routes selected should avoid concentrations of friendly ADA units.

(4) During an aerial crossing of the FEBA by a ranger force, the regimental headquarters may assign a liaison officer to the headquarters controlling the ADA fires in the crossing area. The LNO is then responsible for close and continual coordination. This ensures that the crossing is made without friendly ADA units engaging the ranger force while it is exiting or entering friendly lines.

(5) During ranger operations, the primary source of air defense fires is shoulder-fired antiaircraft missiles carried by specially trained individual rangers. These weapons provide the ranger force commander a limited defense against attack by enemy aircraft. The decision to take these weapons into the objective area depends on the factors of METT-T. Ranger operations should be completed before the enemy can react with ground forces. However, enemy aircraft may be able to react in time to threaten the ranger force. These lightweight missiles offer a highly effective, short-range counter to enemy air attack.

(6) The ranger force commander must define the weapons control status and rules of engagement for all short-range air defense (SHORAD) teams under his control. Priorities of target engagement must also be determined to prevent multiple engagements of a single enemy aircraft. The ranger battalion FSO and the attached ADA officer are responsible for the coordination of ADA fires within the objective area. Standard ADA weapons-control status and rules of engagement may be modified by the ranger force command.

7-5. Aviation support.

a. **Planning.**

(1) The ranger regiment may need extensive aviation support on some missions. The ranger battalion is organized and trained to move quickly to the objective area by air and to be resupplied by airdrop for short periods.

(2) Often the decision to execute a mission depends on the amount and type of available insertion and extraction aviation support. Normally, the theater commander or JTF commander allocates the type and amount of aviation assets to support a ranger operation. The theater or JTF commander's staff is responsible for the planning and coordination concerning aviation support. The ranger force commander must play an active part to ensure mission success.

(3) Due to the extensive training requirements and specialized operational techniques needed to support special operations, special air operations units in the Army and the USAF have been equipped and trained for the mission. These forces train with the ranger regiment and jointly develop tactics, techniques, and procedures of employment. These special units operate diverse aircraft that have unique capabilities. These help the ranger force commander to effectively employ his units. Command of these special units is maintained at a very high level in the chain of command. They are assigned to support a specific operation. Then they return to their parent unit's control.

b. **Coordination.** In addition to the staff of the theater commander or JTF commander, the coordination of aviation support occurs at all levels in the ranger force. These operations often are conducted in phases. A separate air mission commander is then designated. The AMC is responsible for the completion of each phase of the aviation mission, as well as the close and continuous coordination with the ranger force, commander. The ranger regimental air operations officer and the movements control officer are responsible for the planning, integrating, and controlling of aviation support.

c. **US Air Force.**

(1) The USAF provides special aviation support to the ranger regiment and other elements of the special operations force. The USAF has elements that are equipped with MC-130 Combat Talon aircraft (see Figure 7-9) and AC-130 Spectre gunships, as well as HH-53 Pave Low, UH-60, and UH-1 helicopters. These aircraft are specially equipped to conduct long-range infiltration, exfiltration, attack, and search-and rescue operations. The USAF also provides C-141B and C-130 aircraft for strategic and tactical airlift of the ranger force. There are specially trained USAF crews that can operate using AWADS and SKE. This gives them an excellent means of inserting the ranger force during reduced visibility.

Figure 7-9. MC-130 Combat Talon aircraft.

(2) The AFFOR commander provides the airlift control element (ALCE) support at the departure airfield. The ranger air operations officer ensures that coordination of aircraft load limits, air movement times, specific load plans, and manifests is conducted in an effective and timely manner. The air operations officer and the movements control officer ensure that a departure airfield control officer (DACO) from the deploying unit is designated to provide face-to-face coordination at the departure airfield.

(3) The USAF normally provides an airborne battle command and control craft. This is a specially equipped C-130-type aircraft that provides communications and data system down-links to ground stations.

(4) The terminal guidance needed for assault landings and air traffic control in the objective area is provided by a specially trained and equipped CCT. This element has habitual training and operations relationships with the ranger units. The USAF is also responsible for the specially trained and equipped weather teams that can be sent into the objective area with the CCT.

(5) The USAF Tactical Air Command (TAC) includes fighter, interceptor, ground attack, electronic countermeasure, early warning, reconnaissance, and airborne command and control squadrons, any of which may be used to support the ranger force. The Tactical Air Command provides the results of aerial reconnaissance to the AFFOR, who then processes it for use by the theater or JTF commander's staff. This reconnaissance may be by low-level photo, imaging infrared, side-looking airborne radar (SLAR), low-light-level television, or overhead photography.

(6) The Military Airlift Command (MAC) equips and trains special long-range SAR aircraft and crews. These aircraft, mainly helicopters, are equipped with sophisticated navigation, observation, and communications equipment, as well as extended-range fuel tanks and in-flight refueling capabilities (see Figure 7-10). The ranger regiment uses these aircraft for certain missions.

Figure 7-10. In-flight refueling.

d. **US Navy and US Marine Corps.** Both the USN and the USMC have units equipped and trained to support ground forces. The type of aviation support normally provided to the ranger regiment is close air support against targets in the objective area, or as suppression missions against enemy air defense installations. Long-range helicopters from the USMC may be used to support an amphibious operation or to insert special teams during ranger operations. The USN also has a search-and-rescue ability that may be used to support the ranger regiment.

e. **Special operations aviation units.**

(1) The US Army has specially equipped and trained SOA units tasked to support the ranger regiment, as well as other elements of the special operations force. These SOA units have a habitual training association with the ranger regiment and use dedicated aviation assets to -

(a) Insert, extract, and resupply the special operations force.

(b) Conduct armed escort, reconnaissance, surveillance, and electronic warfare in support of the special operations force missions.

(c) Provide airborne command, control, and communications for special operations force elements.

(d) Provide general support aviation during peacetime and contingency operations.

(2) Penetration of enemy airspace is normally done during limited visibility at very low altitudes. Penetrations are planned to maximize the use of existing terrain and exploit blind spots in enemy air defenses. On-board navigational aids are crucial to mission success. There normally is not any terminal guidance to the selected LZ or DZ. Deception operations of a tactical or operational nature help SOA units penetrate to the objective and reenter friendly lines. Advanced helicopters and short-takeoff-and-landing aircraft provide the help needed to do these missions.

(3) To survive enemy air defense systems and prevent mission compromise, SOA units avoid detection. Detected aircraft may reveal the location of supported special operations force ground forces or may be destroyed by enemy ADA fires. The SOA normally avoids detection by operating at night without illumination. This condition requires flight using sophisticated night vision devices. The SOA also operates in other conditions of reduced

visibility, such as clouds, fog, rain, snow, smoke, and dust storms.

(4) Before an operation, the ranger force commander may need to place reconnaissance teams near the objective. Army SOA assets can be used to transport ranger reconnaissance teams.

(5) Elements of SOA provide the ranger regiment with short and intermediate range penetration, resupply, and extraction capabilities. Long-range aircraft must be provided from USAF special operations assets.

(6) The ranger regimental or battalion commander is normally provided a liaison aircraft while the ranger force is in the REMAB. This ensures that he can travel quickly to the dispersed headquarters of the agencies providing support for the mission. This aircraft and crew are provided by the theater or JTF commander from EAC general support aviation assets. It may be a helicopter or fixed-wing aircraft, depending on the location of the REMAB to a suitable landing field.

(7) Army helicopters and fixed-wing aircraft may be used to provide SEMA aerial reconnaissance; radar, infrared and photographic imagery; and electronic intelligence support to the ranger force commander. This support is normally provided by the aerial exploitation battalion of the theater Army military intelligence group.

7-6. Engineer support.

a. The ranger regimental and battalion operations officers plan and request engineer support for ranger operations.

b. The most common engineer-type tasks needed during a ranger operation are the destruction of enemy facilities, equipment, and material by use of conventional explosives; and the interdiction of major lines of communication. To reduce the need for external support, the ranger regiment trains selected individuals in advanced demolition techniques. All members of the ranger regiment are well trained in the basic employment of conventional demolitions. Therefore, there is no need to attach combat engineers to conduct normal demolition missions. If specialized skill or knowledge is needed for a specific mission, qualified personnel can be attached.

c. During the occupation of the REMAB and the rehearsal for an operation, the ranger regiment needs construction engineer support to build simulated target areas. The simulated target areas should be camouflaged to prevent the enemy from recognizing the target. The construction engineers augment the ranger unit's self-help construction efforts. Limited-facilities construction and improvement may also be needed at the REMAB to provide security and isolation. The theater commander or JTF commander provides the construction engineer support.

d. Certain missions, such as those of interdiction, may need engineer special weapons teams. The teams are provided by the theater commander or JTF commander, or they may come from other special operations force assets. They are attached to the ranger force for a specific mission and released at the completion of the extraction and debriefing phase.

e. Some operations may need the ranger force to use captured enemy heavy construction equipment. Selected members of the ranger regiment are trained to operate such equipment.

However, construction engineer personnel may be attached to the ranger force for this mission.

CHAPTER 8

Combat Service Support

Combat service support consists of the logistical and administrative effort to maintain the battalion's ability to fight. The commander forecasts his needs for CSS. He employs CSS units to ensure accomplishment of the mission.

8-1. Ranger support element.

a. The ranger support element is a non-TOE organization formed from elements of CSS units collocated with CONUS-based ranger units augmented by selected personnel from the host installation. The mission of the RSE is to provide all the CSS needed to outload and deploy a ranger unit from CONUS. The personnel designated as part of the RSE have the same alert notification, reaction, and security requirements as members of the ranger unit they support. The RSE is normally built around a command and control group provided by a major CSS unit on the ranger unit's host installation. The remaining CSS elements and teams are then designated. An RSE consists of--

(1) An emergency operations center liaison group that provides timely coordination between the ranger unit and the functional elements of the RSE.

(2) A command and control group that coordinates and directs the actions of the functional elements.

(3) A supply section responsible for the timely issue of all Class II, III, IV, and VII items. This section maintains a stockage of contingency items marked for use by the resident ranger unit. It also issues items out of common stocks.

(4) A transportation section moves equipment and personnel from the unit area to the departure airfield and picks up equipment from local storage sites. It may also operate material-handling equipment and trailers to support loading of aircraft.

(5) An ammunition section responsible for the requisition, storage, issue, loading, and security of the ranger unit's basic load(s) of conventional munitions.

(6) A mess section provides all Class I support to the ranger unit. This includes operating a dining facility, providing food to remote sites, and drawing and issuing operational rations.

(7) A load team loads and unloads all equipment and supplies for the ranger unit. This team prepares pallets for loading, documents hazardous or dangerous cargo, and loads equipment pallets on USAF aircraft. It can also provide limited security for loads.

(8) A communications and electronics section provides secure communications among the RSE subelements, ranger unit, EOC, and DACG.

(9) An engineer platoon that is responsible for the construction of targets and rehearsal areas. This element of the RSE may or may not be activated, depending on the mission given the ranger unit. If the ranger unit deploys from CONUS to the objective area, this platoon has a key role to play in the premission preparation.

(10) A maintenance team whose primary mission is to ensure that the vehicles used by the RSE remain operational. It also provides predeployment maintenance and Class IX support for ranger unit vehicles and equipment.

(11) A medical team that provides unit-level medical care while the organic ranger medical section completes its outloading preparations. This team also provides any Class VIII supplies that the ranger unit may need.

(12) A security team responsible for security at the EOC, ranger unit location, and departure airfield. This team enforces stringent physical security precautions and maintains a constant counterintelligence (CI) surveillance of the outloading procedure. The security team provides armed guards for the protection of ammunition convoys, equipment movement, and vehicles.

(13) A rigger detachment that maintains, packs, rotates, and issues the personnel and equipment parachutes required by the ranger unit. It maintains enough stock of personnel parachutes to conduct a battalion-level parachute assault. The riggers lend technical skill to the rigging and aerial delivery of bulky, unusual, or delicate equipment.

b. The primary mission of the RSE is to provide the support necessary to quickly load and deploy the ranger units. The theater commander provides CSS to the ranger force once it has arrived at the OCONUS REMAB or ISB. The transition of support from the RSE to the CSS assets of the theater commander may be delayed for certain subelements in order to ensure a smooth deployment and enhance mission accomplishment. These subelements are released when the in-theater CSS assets can provide the support. Retained subelements of the RSE may be released from support requirements all at once, released in increments, directed to continue to provide support until mission completion, or augmented by in-theater CSS assets. The maintenance, medical, security, and rigger elements are those most likely to continue providing support OCONUS. This is likely if deployment is to an austere ISB pending immediate employment.

c. Once a ranger unit has been deployed OCONUS and has closed to a REMAB or ISB, the theater commander's staff coordinates CSS. If 1st SOCOM elements have deployed to the theater or operational area, the special operations support battalion (SOSB) of the 1st SOCOM coordinates CSS. Ranger logisticians at regiment and battalion level are mainly planners with limited logistics operations ability. Ranger logistical and personnel officers coordinate with CSS elements who provide such services.

d. The Theater Army Materiel Management Center (TAMMC) is the point of contact for information concerning in-theater logistical assets. In a developed theater of operations, a series of theater Army support groups will have been set up. One of these groups is assigned the mission of supporting the ranger force. The 1st SOCOM may also deploy certain assets available to provide CSS to the ranger force. The many CSS sources, the rapidly changing nature of SOF operations, and the unique CSS requirements of ranger units combine to make providing logistical and personnel support a challenge. The ranger unit's logistics officer must be included in the early phases of planning and deployment.

8-2. Logistics and personnel.

a. The echelons-above-corps CSS structure provides support to the ranger regiment as directed by the theater Army (TA) commander. In a low-intensity contingency involving a corps or less, there may be no echelons-above-corps logistics and personnel support elements. The corps support command (COSCOM) involved may be able to meet all needs from its own sources. However, in a combined operation in a high-intensity environment, the theater commander assembles a large CSS organization to provide support to the TA. This CSS base normally provides for all the ranger unit's requirements. During contingency operations in an area where there is no TA, the JTF commander provides the CSS elements. Until the buildup of these elements is complete, the ranger regiment must be prepared to operate from austere facilities with only the supplies it brings from CONUS.

b. The ranger regiment is normally assigned to the theater command for CSS. The TA commander, as directed by the theater commander, provides CSS to the ranger regiment and sets up support priorities. The EAC CSS organizations are under the command of the TA commander and consist of subordinate area and functional commands. The TA plans, coordinates, and executes its CSS functions through area-oriented support commands and specialized subordinate organizations. During the early stages of conflict, the TA may be organized as shown in Figure 8-1.

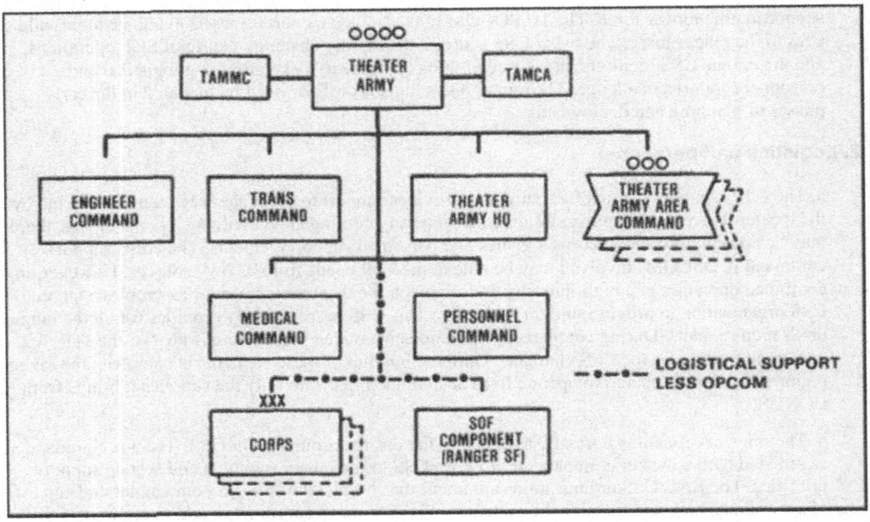

Figure 8-1. Ranger force relationship to theater Army.

c. The Theater Army Area Command (TAACOM) is the key CSS operator in the communications zone. It functions as a major subordinate command under the theater Army. The primary mission of the TAACOM is to provide direct support CSS to units such as the ranger regiment located in or passing through its assigned area. While located in an OCONUS REMAB, the ranger regiment receives most of its logistical support from the TAACOM.

d. The TAACOM area support group (ASG), which is subordinate to the TAACOM, is structured with operating units to provide direct support CSS to units located in or passing through its assigned geographical area of responsibility (see Figure 8-2). Area support groups are usually located along the lines of communication. The ASG is the prime source of CSS for the ranger regiment. It contains the units that actually provide services requested by the ranger unit logistics officer.

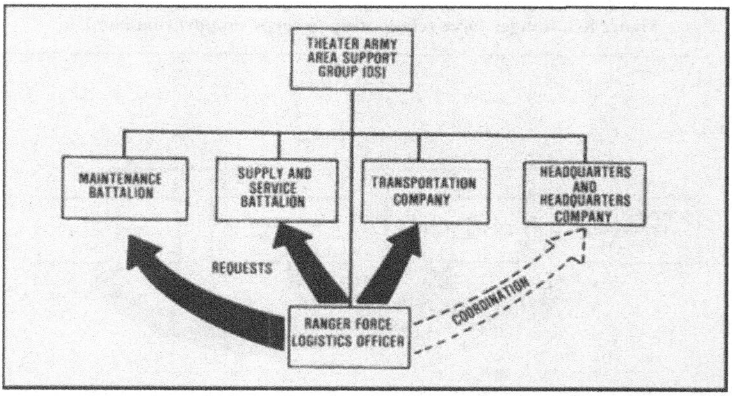

Figure 8-2. Ranger force relationship to theater Army area support group.

e. The ranger regiment is not normally assigned to a corps for employment. However, it may be placed under a corps' operational command for a specific mission. The ranger unit logistics and personnel officers must coordinate with the COSCOM that would then be providing CSS. (See Figure 8-3.) Corps support groups are major subordinate elements of the COSCOM. They provide command and control of COSCOM companies and battalions, which provide direct support (DS) and general support (GS) supply and field services to nondivisional units, such as the ranger battalion. (See Figure 8-4.)

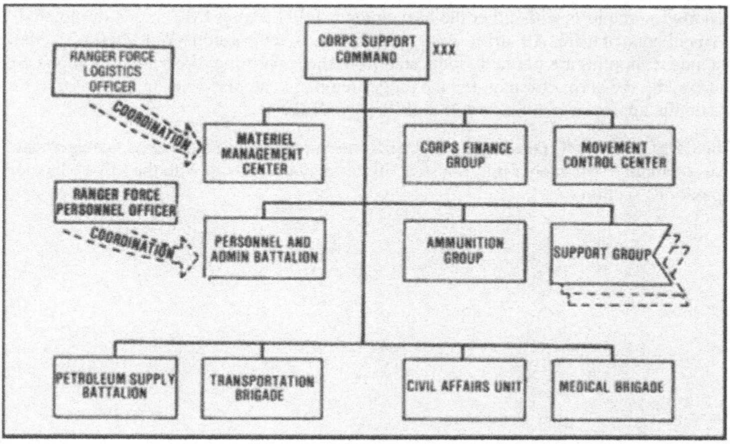

Figure 8-3. Ranger force relationship to corps support command.

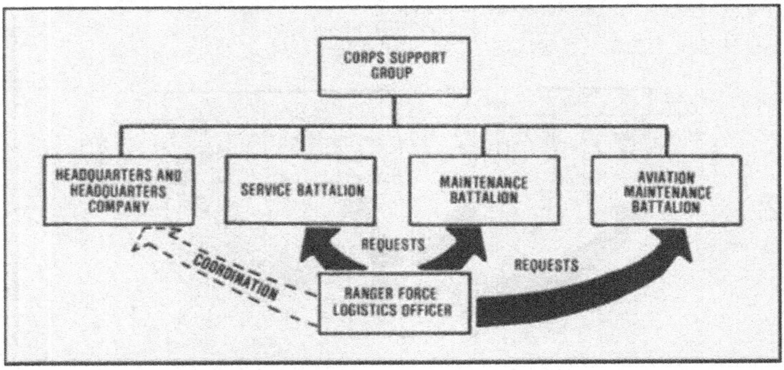

Figure 8-4. Ranger force relationship to corps support group.

f. Normally, the ranger regiment receives its logistical support from the corps support group responsible for the geographical area in which the REMAB or ISB is located. The corps support group provides all supplies except Classes V and VIII. The medical brigade provides Class VIII supply and medical maintenance support. The ammunition group provides Class V support. Financial and personnel services are provided by the corps finance group and the personnel and administration battalion, respectively.

g. Transportation needs are met by the transportation brigade. The ranger unit places ground transportation requests with either the Movement Control Center (MCC) or a designated movement control team. All airlift supply requirements are passed IAW FM 100-27. Standard and nonstandard supplies are normally requested from the supporting direct support unit (DSU). If authorized by the corps commander, the ranger regiment can place requirements for nonstandard items on the appropriate division within the corps MMC.

h. The COSCOM MMC (see Figure 8-5) performs supply and maintenance management for all classes of supply. The ranger unit logistics officer may coordinate with the MMC division chiefs or operations section.

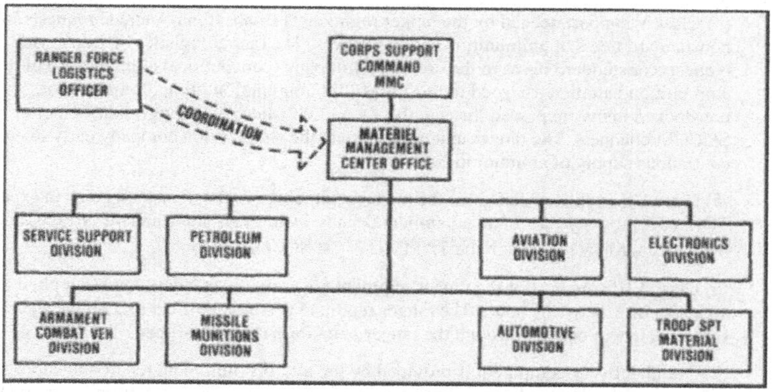

Figure 8-5. Ranger force relationship to COSCOM MMC.

i. Critical logistics operations, such as those supporting ranger and other SOF elements, need intensive management. They may be assigned to special staff sections that act as special operations force logistics managers. If designated, a logistics manager for special operations force elements would be a major point of contact for logistical support for the ranger regiment.

j. The ranger regiment may also receive logistical support from the SOSB. Normally, the SOSB provides administrative and logistical support to the Headquarters, Army Special Operations Forces (ARSOF). It complements the support provided by the Army component in the theater of operations by coordinating, augmenting, or assisting CSS forces. The SOSB provides unique quick-reaction support to the ARSOF or deploying SOF elements.

k. The ranger regiment does not make inordinate demands on the logistical system. However, a ranger unit may need certain nonstandard or low-density equipment. The concurrent need for nonstandard Class IX items requires management by logisticians to ensure support.

(1) Ranger battalion mess sections are responsible for Class I operations at the REMAB. The ranger regimental headquarters, which has no mess section, either collocates with one of the battalions or is supported by a mess facility operated by a nearby unit. It may share this facility on an area basis. The ranger unit personnel officer coordinates with the supporting DSU, which initiates the issue of Class I supplies.

(2) Class II, IV, and IX support to the ranger force is normally not of a great volume. It may consist of some nonstandard items. Depending on the need, the item, and the security, the SOSB may get some nonstandard items directly.

(3) Class III support needed by the ranger regiment is small and limited to standard items. The normal source of supply for Class III is the supply and service company providing area support.

(4) Class V support needed by the ranger regiment is small. It may include requests for nonstandard items of ammunition and explosives. The ranger logistics officer passes requests for standard items to the ordnance company (conventional ammunition) of the ammunition battalion assigned to the TAACOM ammunition group. Requests for nonstocked items are passed through the TAACOM Materiel Management Center or 1st SOCOM channels. The ranger unit must contact the ammunition company early to ensure a continuous supply of ammunition.

(5) Class VII supplies needed for the ranger regiment are often nonstandard or in excess of TOE. (See Appendix C for an equipment density listing.) All nonstandard or excess requests must be coordinated through the TAMMC or 1st SOCOM channels.

(6) Class VIII supplies for the ranger regiment are normally standard items supplied by the area support Class VIII point. The ranger regiment personnel officer sets up the Class VIII supply points of contact through the ranger battalion's medical officer.

(7) Aerial delivery equipment is provided by the airdrop equipment repair and supply company of the theater Army support group (TASG) (GS). There is normally one such company for each TAACOM. This unit supplies and repairs aerial delivery equipment. It does not pack and rig supplies for aerial delivery. The ranger regiment's requirements for aerial delivery supplies normally consist of standard items. Certain low-density items, such as HALO or HAHO parachute equipment, may also be provided through 1st SOCOM channels.

(8) The supporting motor transportation group of the transportation command is responsible for unit movement of the ranger regiment. The ranger regiment does not have any organic transportation assets. It depends on outside support whenever motor movement is needed.

8-3. Resupply.

a. Resupply operations for ranger units are normally covered during the planning and coordination phase of the mission. Ranger units try to complete their mission quickly and then be extracted or exfiltrate from the objective area. Ranger units carry into the objective area the equipment and munitions they need. They do not normally remain long enough to need extensive resupply. Certain operations may require the use of bulky or heavy items of equipment that cannot be carried by personnel. If the ranger force is airlanded on or near the objective, one or more of the aircraft may contain palletized supplies. These can be quickly off-loaded for future use.

b. If the ranger force is inserted into the objective area by parachute assault, aircraft can also drop resupply loads either just before or just after the personnel drop. These loads might include extra water, ammunition, demolitions, or barrier material. During peacetime, preloaded airdrop pallets are stored at CONUS depots for such contingency use. These mixed loads can be quickly moved to a departure airfield and inserted along with, or following, the ranger force. By dropping a resupply of water, food, equipment, and munitions with the ranger force insertion, the heavy-drop aircraft do not have to conduct a separate penetration of the enemy air defense zone. The ECM and SEAD programs used in the insertion of the personnel aircraft mask the penetration of the heavy-drop aircraft.

c. The aircraft dropping ranger personnel can also carry door bundles to deliver bulky or heavy items to the objective area. These door bundles, normally made up of A-7 and A-21 containers, may be color-coded to identify different loads or units. The most common door bundle loads are such items as 90-mm recoilless rifles and ammunition, 60-mm mortar ammunition, Dragon missiles, SHORAD missiles, or more demolition material. These bundles are pushed from the personnel aircraft just before the lead jumper's exit. They may be marked with lights or infrared sources for locating them at night.

d. If the tactical situation requires that a ranger force receive a resupply during the course of an operation, the most common method would be by airdrop. Airdrop could be needed if the operation takes longer than planned, initial supplies are lost, or changes occur in the operational plan. The ranger unit logistics officer and operations officers normally plan for resupply needs and schedule follow up loads of Classes I, II, IV, V, and VIII. There are two different airdrop requests channels for the ranger force to use:

(1) Nonemergency airdrop requests are sent through logistics channels. The ranger LOC sends requests for the airdrop of supplies and equipment to its supporting supply unit, and the supporting unit forwards the requests to the COSCOM or TAACOM MMC. The MMC coordinates with the movement control center (MCC). Jointly they direct the actions of the supply and transportation units. The MCC sends the request to the joint transportation board (JTB). The board validates it and assigns it a priority. The JTB then sends the request to the airlift control center for action. If the items requested are not on hand in the airdrop company, the MMC directs the proper supply activity to prepare the items for shipment to the airdrop unit. The MMC coordinates with the MCC, and the MCC ships the items. The airdrop unit prepares the supplies for the mode of delivery to be used. It also makes sure the supplies are delivered to the air terminal at the right time. (See Figure 8-6.)

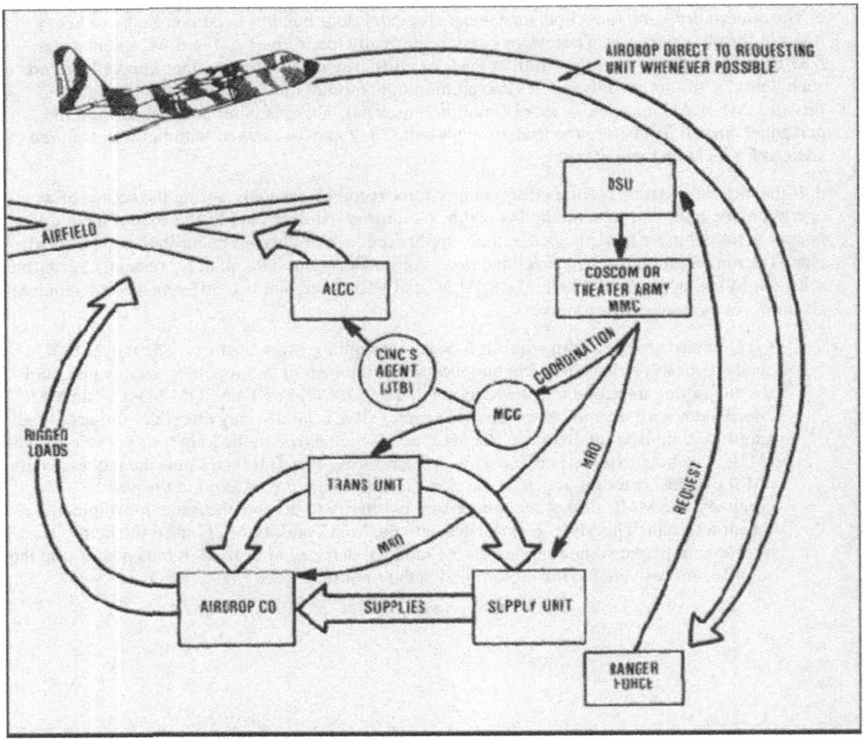

Figure 8-6. Ranger requests for nonemergency airdrop.

(2) Emergency requests for airdrop are passed through command channels in the quickest, most secure way.

(a) The request is usually sent to the command level that has authority to approve it, usually the tactical operations center at corps or above. The Air Force airlift officer at the direct air support center (DASC) coordinates with the Assistant Chief of Staff, Operations (G3) at the tactical operations center to determine whether airlift resources can be used for the mission.

(b) The TOC sends a request for an immediate airlift to the joint force commander's (JFC) designated agent for validation. At the same time, the airlift liaison officer advises the airlift control center (ALCC) that a request is coming through channels so plans can be made. When the ALCC receives the validated request from the JFC agent, it diverts or cancels less essential missions and provides airlift support. The TOC sends the request for supplies or equipment to the MMC.

(c) The MMC and MCC jointly direct the actions of the supply and transportation elements. The airdrop company stocks Classes I, III, and V supplies. It holds these supplies for emergency missions. If the supplies and equipment requested are not on hand in the airdrop company, the MMC directs the proper supply activity to prepare the items for shipment to the airdrop unit. The MMC coordinates with the MCC to ship the items. The airlift is arranged by the ALCC.

(d) The airdrop unit prepares the supplies according to the mode of delivery to be used. It coordinates with the MCC and delivers the supplies to the designated point at the air terminal (usually the cargo aircraft ramp). (See Figure 8-7.)

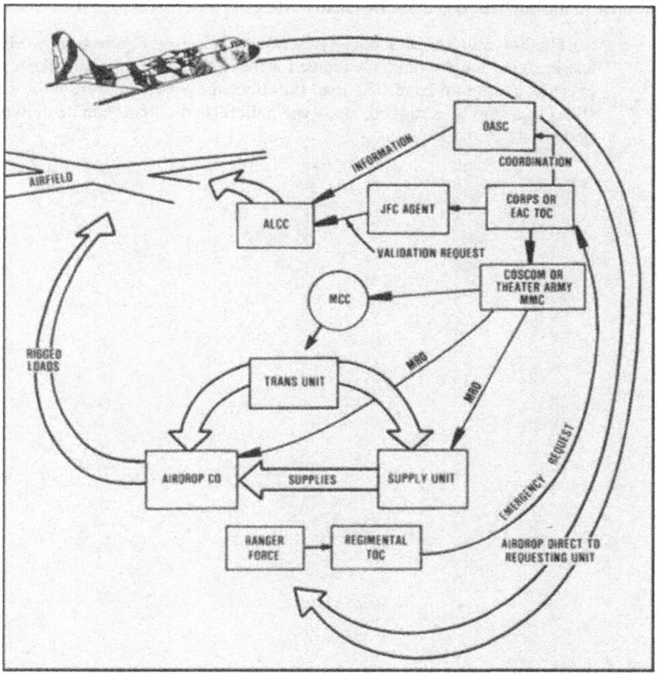

Figure 8-7. Request channels for emergency airdrop.

e. Aircraft delivering airdrop resupply deep behind enemy lines need to take precautions to avoid enemy detection and antiaircraft fire. The USAF is responsible for most of these precautions, such as ECM, suppression of enemy air defense (SEAD), and WILD WEASAL support. These do not concern the ranger commander. The choice of the type of parachute delivery system to use, however, may affect the ground commander's plan. The ranger force commander must specify any tactical restrictions that would require a specific type of delivery system. The ranger LOC and regimental logistics officer must advise the ranger force commander on the type of parachute system to be used. Conflicts must be quickly resolved so as not to interfere with the tactical operations.

(1) The safest way for the airdrop aircraft to penetrate enemy air defense and remain undetected is often by use of very low-level flight. There are four low-level parachute delivery systems that can be used without the aircraft being required to climb to a higher drop altitude and risk detection by enemy ADA.

(a) The low-altitude parachute extraction system (see Figure 8-8) needs a length of level ground with enough clearance for the delivery aircraft to make a very low pass over the extraction zone. The load is extracted by parachute from the aircraft and slides to a stop on a rugged, skid-type pallet. Heavy loads can be delivered using this system.

Figure 8-8. Low-altitude parachute extraction system.

(a) The containerized delivery system (CDS) (see Figure 8-9) consists of multiple, individually rigged A-22 containers, each with its own parachute, weighing up to 2,000 pounds each. The system provides single-pass delivery of up to 16 containers by C-130 aircraft and up to 28 containers by C-141 aircraft. The CDS loads may be delivered into drop zones using multiple points of impact to allow for tactical separation. The ranger force commander can direct the use of multiple points of impact, noting the advantages to be gained against the requirement for multiple passes.

Figure 8-9. Containerized delivery system.

(c) The high-speed, low-level, airdrop system (HSLLADS) (see Figure 8-10) is a single A-21 container specially rigged to withstand the parachute opening shock when airdropped from C-130 aircraft at high speed. This system can be used to deliver up to 600 pounds for each container, with a maximum of four containers for each pass, at speeds up to 250 knots.

Figure 8-10. High-speed, low-level, airdrop system.

(d) The CTU-2/A is a high-speed, aerial-delivery container (see Figure 8-11) that can be used to deliver supplies from high-performance aircraft flying at a minimum altitude of 300 feet above ground level and a maximum airspeed of 425 knots. The CTU-2/A is carried on the bomb racks of fighter or bomber-type aircraft. Upon release, a pilot parachute deploys the main chute and the container descends slowly.

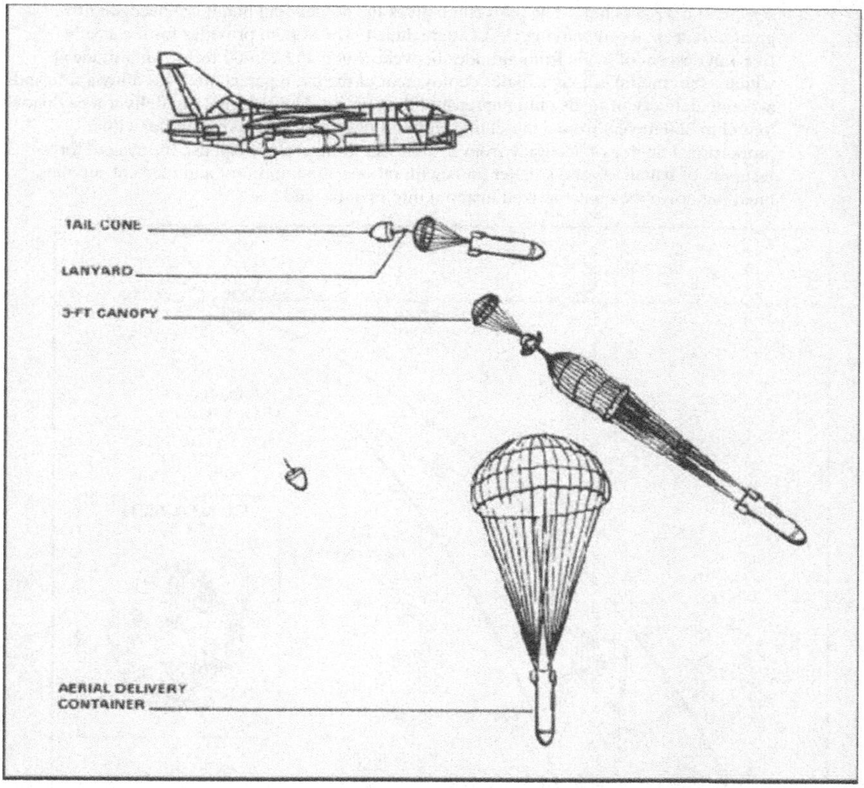

Figure 8-11. CTU-2/A high-speed, aerial-delivery container.

The CTU-2/A can be used to deliver up to 500 pounds of supplies such as weapons, water, food, or munitions. The container can be destroyed by burning. The main advantage of this system is that it can be delivered by high-performance aircraft deep behind enemy lines and in a dense air defense environment. The accuracy of this system is equal to that of a conventional bomb strike.

(2) Sometimes it may be better to drop resupply loads to the ranger force from a high altitude. The HAARS (see Figure 8-12) permits containerized unit loads weighing from 200 to 2,000 pounds to be delivered from aircraft at speeds up to 150 knots from up to 25,000 feet AGL. The HAARS consists of a cargo parachute, an airdrop container, an altitude

sensor, and a pilot chute. The pilot chute gives the descending bundle a speed slightly greater than an accompanying HALO parachutist. The system provides for the steady free-fall descent of loads from altitudes between 2,000 and 25,000 feet to an altitude at which a barometric sensor actuates deployment of the main parachute. This allows safe and accurate delivery of loads onto unprepared drop zones. The HAARS can deliver a payload to within 260 meters from a target impact point from a 10,000-foot altitude, with a proportional degree of accuracy from 25,000 feet. Ranger units can use the system for resupply of battalions and smaller units with rations, ammunitions, and medical supplies, breaking down the containerized material into man-packed loads.

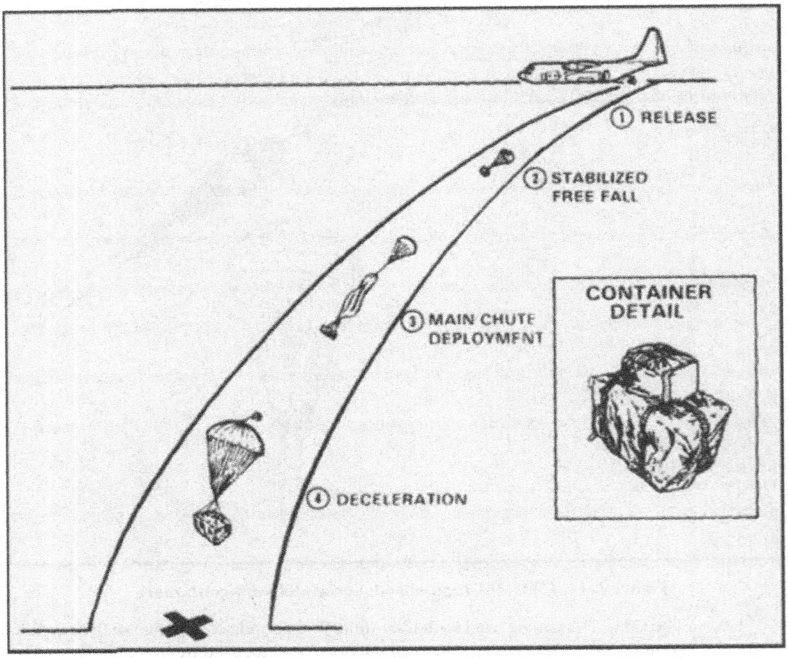

Figure 8-12. High-altitude, airdrop resupply system.

(3) The ranger force may need aircraft loads delivered during periods of poor visibility. The adverse weather aerial delivery system allows aircraft to drop resupply bundles accurately in adverse weather or darkness. These AWADS operations can be conducted safely and effectively with a minimum 300-foot ceiling AGL and a minimum visibility of less than half a mile. The minimum-drop altitude is 500 feet above the highest obstruction. These minimum ceiling or visibility restrictions do not apply to the Combat Talon aircraft.

f. Resupply to the ranger force can be provided by Army helicopters (see Figure 8-13). Normally this is done by SOA units. The danger to helicopters during long-range, low-level penetration of enemy rear areas requires that resupply be conducted when there is limited visibility.

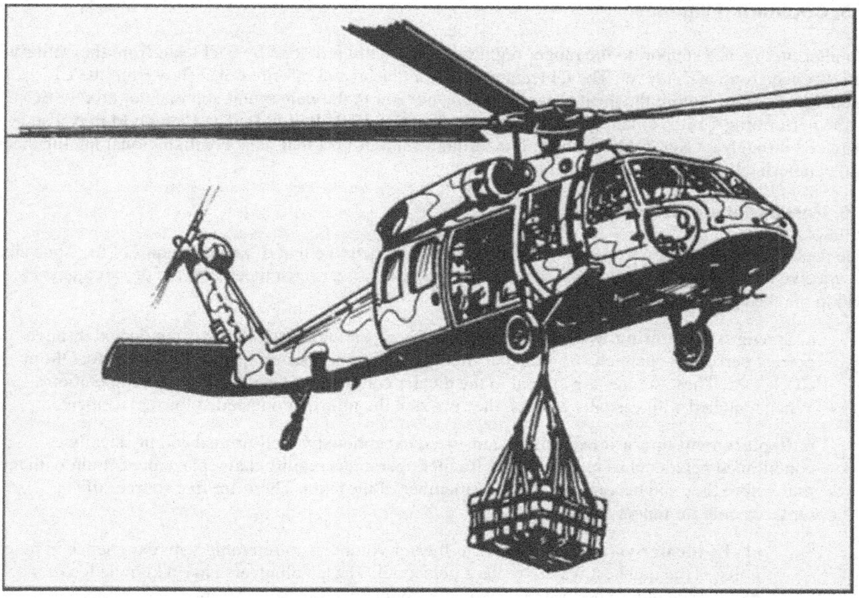

Figure 8-13. Resupply bt helicopter.

8-4. Military police.

a. The ranger regiment needs military police support to provide security at the REMAB. Although ranger units can provide their own security and local defense, using military police for this purpose allows the ranger unit to concentrate its manpower and training on mission preparation. A military police security company is normally designated to provide support to 1st SOCOM elements located in the communications zone. Military police provide security patrols of the REMAB, convoy security, and vehicular search. They provide entry and exit control of the sensitive compartmented information facility (SCIF).

b. A preemptive strike by enemy special operations forces is a threat to the ranger regiment while located in the REMAB; either by conventional attack or by terrorist-type bombing. The preparation and location of the headquarters, the type of buildings, the location of landing zones, the traffic flow, and the size of the local population all influence the size of the military police unit. A military police platoon, augmented with bomb and guard dog teams, should be enough security at the REMAB. These security forces would not normally deploy from the REMAB to an ISB with the ranger unit.

8-5. Counterintelligence.

Counterintelligence support to the ranger regiment is normally provided by a CI team from the military intelligence group at TA level. The CI team provides a liaison and informational flow from the CI agencies operating within the theater. It also acts as advisor to the ranger unit commander on OPSEC while performing CI surveillance operations in the vicinity of the REMAB. The CI support may also be provided through 1st SOCOM channels. It is through 1st SOCOM that liaison with national intelligence and counterintelligence agencies is achieved.

8-6. Personnel service support.

The personnel service support needs of a ranger unit are much like that of any other unit of the same size. To receive the support his unit needs, the ranger personnel officer must meet with the proper agencies within the theater of operations.

a. **Strength accounting.** Within the ranger regiment, strength accounting is conducted through normal personnel channels. Regimental strength figures are passed quickly to the TAACOM or COSCOM. These figures are crucial to the theater commander for planning future operations. When matched with casualty reports, they provide the information needed for replacements.

b. **Replacement operations.** Ranger units need exceptionally well-trained and physically conditioned personnel as replacements. Ranger operations require that replacements train with the unit before they can become an effective member of the team. There are five sources of replacements for ranger units:

(1) **In-theater volunteers.** Use of in-theater volunteers, preferably with experience in ranger units, is the quickest way to replace personnel. These volunteers have a knowledge of the basic skills required of rangers and have some operational experience. The ranger regimental personnel officer coordinates with the theater personnel officer to screen and select those volunteers with the required MOSs and grades. Volunteers from other special operations force components, or from other theaters of war, are identified by DA and screened for assignment to the ranger regiment. Even though these volunteers have individual training and experience, it takes several weeks to become fully integrated into the ranger unit's operations and to achieve the needed skill level.

(2) **Reorganization of units.** This method of reconstitution is used when two or more ranger units have high personnel losses. The remaining members of these units are reorganized into a single unit. This method provides the ranger regimental commander with enough ready-trained, physically fit personnel to form a full-strength unit. Extensive unit training must be conducted before such a reorganization regains the unity and esprit of its component parts. The high standards found among the subelements of the ranger regiment makes this a good method of reconstituting a ranger unit. The units that have lost all their personnel may be reconstituted later, using some other method, or left unreconstituted,

depending on the situation.

(3) **Hospital returnees.** These personnel, former members of the ranger regiment who were injured, evacuated, treated, and have now recovered, are valuable assets. They provide the unit with a major reconstitution ability. These personnel should be returned quickly to their original battalion. This enhances unity and increases morale and esprit. Returnees may not be physically able to return to full duty right away. While recuperating, they may be assigned to limited duty as instructors. The ranger unit uses the returnee's combat experience and individual skills to screen and train newly assigned volunteers. Lessons learned in combat can help prepare the unit for future operations.

(4) **Replacement volunteers from CONUS.** These personnel have training in the basic skills and may have completed the ranger course, but they do not have any ranger unit experience. They need several weeks of unit training to become prepared for combat operations.

(5) **Ranger indoctrination program.** During wartime, this program provides screening and training of newly assigned personnel. Volunteers who have neither served in a ranger unit before nor have graduated from the ranger course require extensive individual training before they are ready to begin unit training with a ranger battalion. This method of reconstitution of ranger units requires the longest training period. During this reconstitution phase, a ranger battalion needs extensive use of ranges, maneuver areas, and support by aviation and other special operations-type units. The ranger regimental commander may choose to move the program from CONUS to a location near the regiment's REMAB.

c. **Casualty reporting.** Casualty reports are handled through regimental personnel channels and must be timely and accurate. The reports are used to check strength accounting data, to assist graves registration personnel in the recovery and identification of remains, and to notify next of kin.

d. **Health service.** The health service support needed by the ranger units deployed OCONUS is usually small. Area medical support provided by host country or theater medical command assets usually meets their needs. Evacuation of wounded members of the ranger force begins upon extraction from the objective area. Personnel wounded in action are evacuated through USAF or Army channels to the nearest medical facility that can provide treatment. The ranger force medical officer and personnel officer coordinate with the area medical support facility providing preventive and routine medical care to the unit while at the REMAB. Vaccine needed for innoculation of the ranger force members and any special Class VIII supplies needed for the mission are drawn through normal medical channels.

e. **Financial service.** Financial support varies with the situation. In mid- and high-intensity conflicts, regular paydays are suspended in the theater of operations. As a nondivisional unit, the ranger regiment coordinates with the theater finance center or corps finance group for financial support. In the REMAB, ranger units are provided financial support from the closest finance support unit. The ranger unit personnel officer appoints Class A agents to help with the disbursement of funds. The ranger force may be given a specific amount of local currency or military scrip depending on the length of time in the REMAB and the operational situation. The ranger unit personnel officer also coordinates with the theater finance center to get any third-country currency that may be included as part of an escape-and-evasion kit provided to ranger elements.

f. **Postal service.** Postal service to the ranger unit is through normal channels. Mail for members of the deployed force is forwarded by the rear detachment commander to an established theater Army general support postal activity. This activity forwards the mail to the postal unit designated to support the REMAB site. The ranger unit commander appoints subordinate unit officers as censors. The exact location of the REMAB, and all aspects of the mission and its location, remains classified until released by an official Army spokesman. The ranger unit commander must ensure that the forwarding of mail, or the writing home of personal letters, does not endanger the ranger force mission. All outgoing mail is censored during wartime. For security reasons, outgoing mail may be held for a short time before it is forwarded. Mail from the ranger unit is never sent by host country or other nation's postal services.

g. **Chaplain activities.** Chaplain activities are provided by the chaplain assigned to the ranger battalion. He provides unit coverage through services, rites, and sacraments regardless of his or the unit member's religious affiliation.

h. **Legal service.** Ranger units do not need extensive legal service support. Legal service support is obtained through coordination with the theater Army staff judge advocate.

APPENDIX A

Organization and Equipment Introduction of the Ranger Regiment

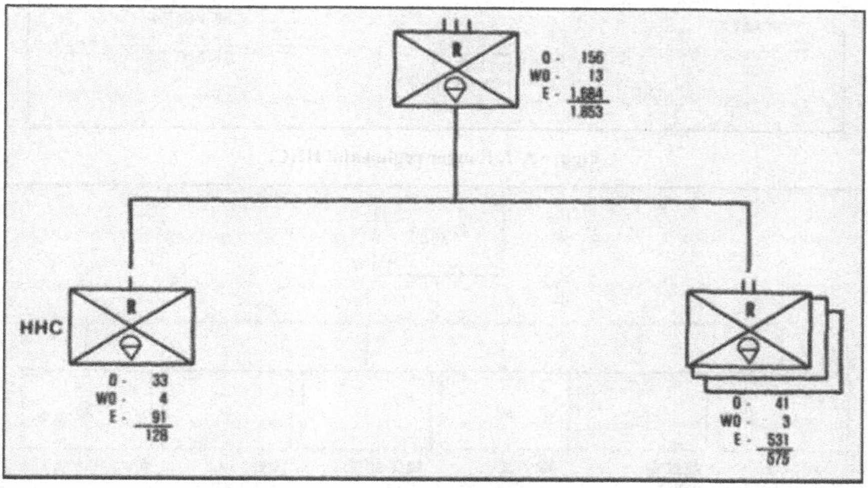

Figure A-1. Ranger regiment.

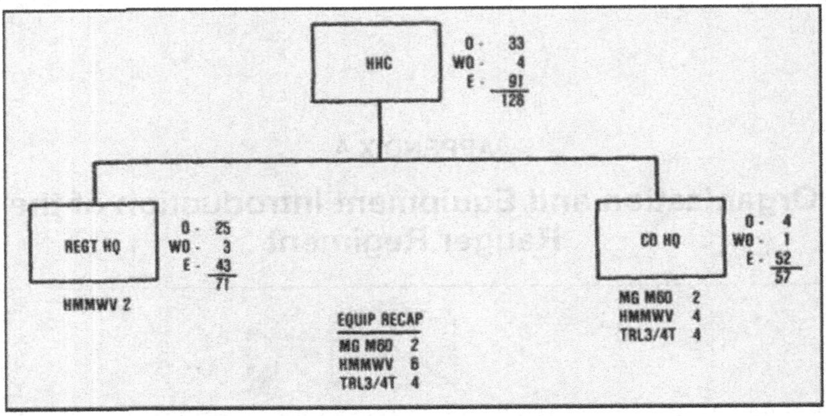

Figure A-2. Ranger regimental HHC.

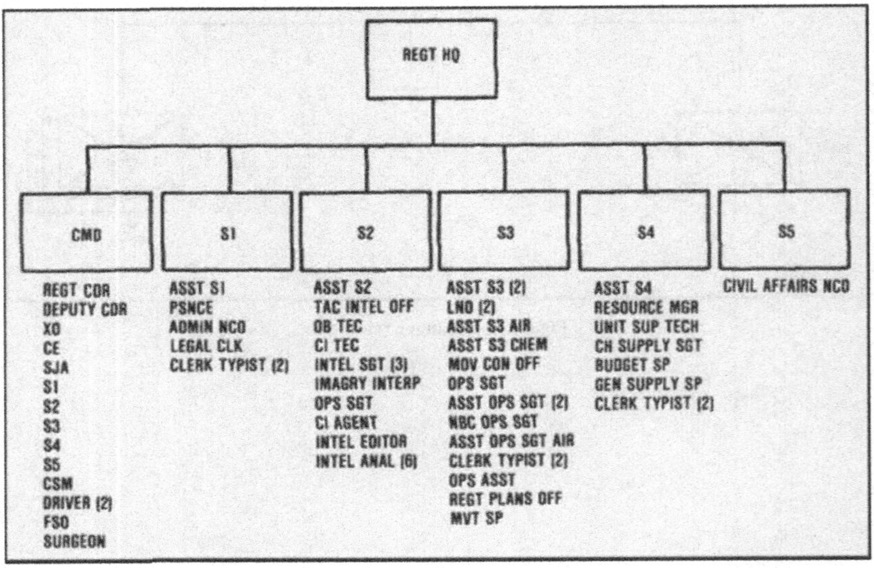

Figure A-3. Ranger regimental staff.

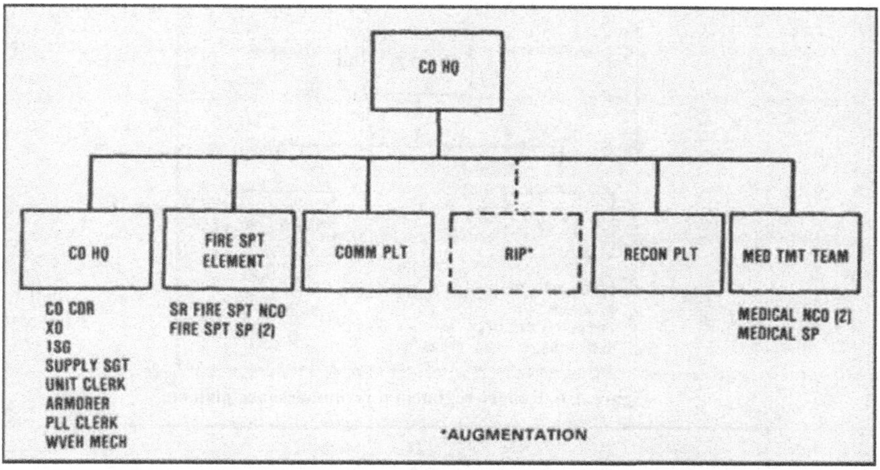

Figure A-4. Ranger regimental headquarters company.

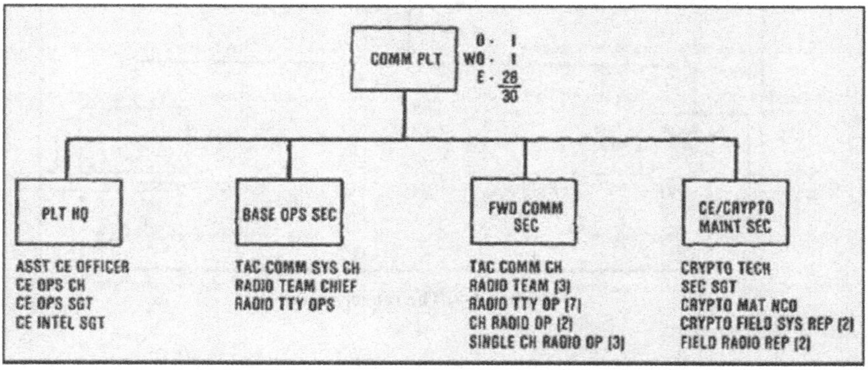

Figure A-5. Ranger regimental communications platoon.

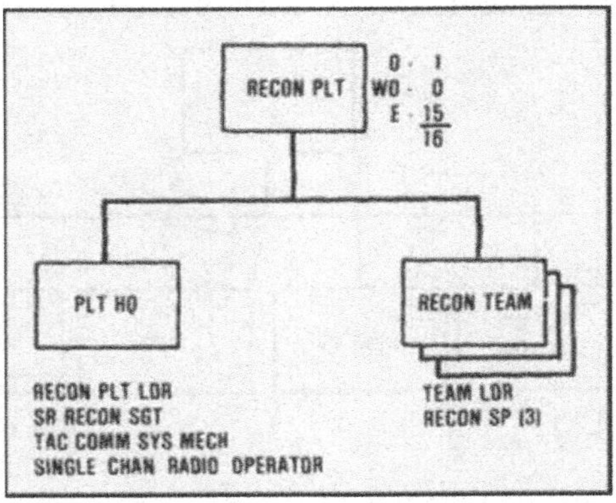

Figure A-6. Ranger regimental reconnaissance platoon.

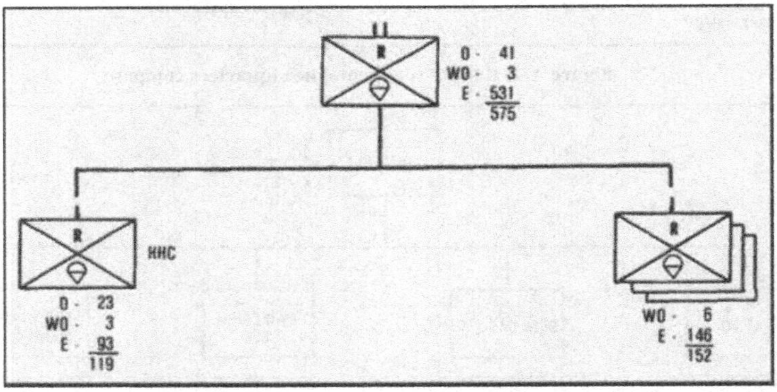

Figure A-7. The ranger battalion.

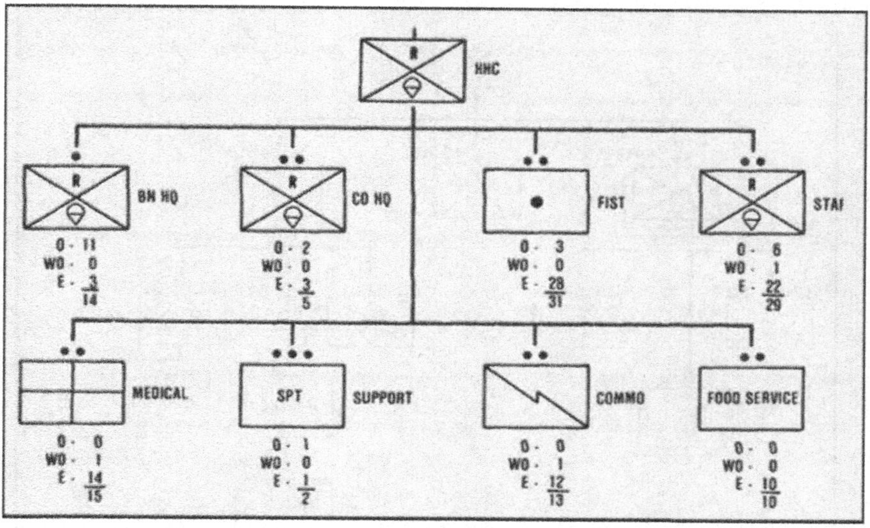

Figure A-8. Ranger battalion HHC.

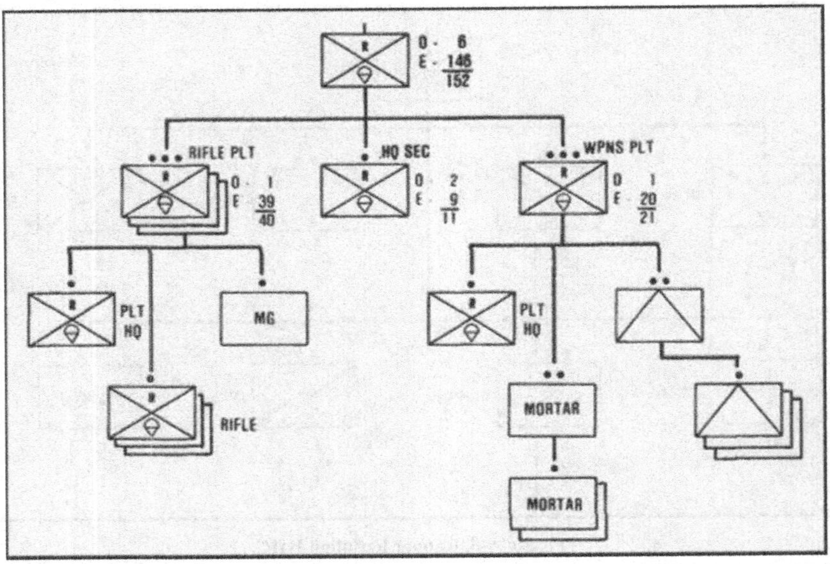

Figure A-9. Ranger rifle company.

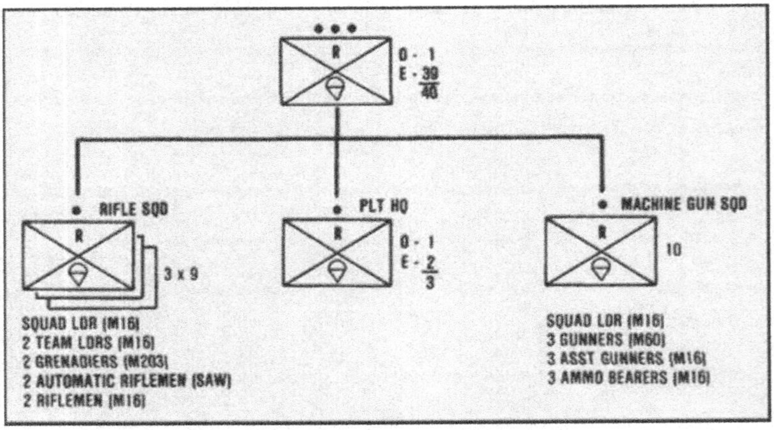

Figure A-10. Ranger rifle platoon.

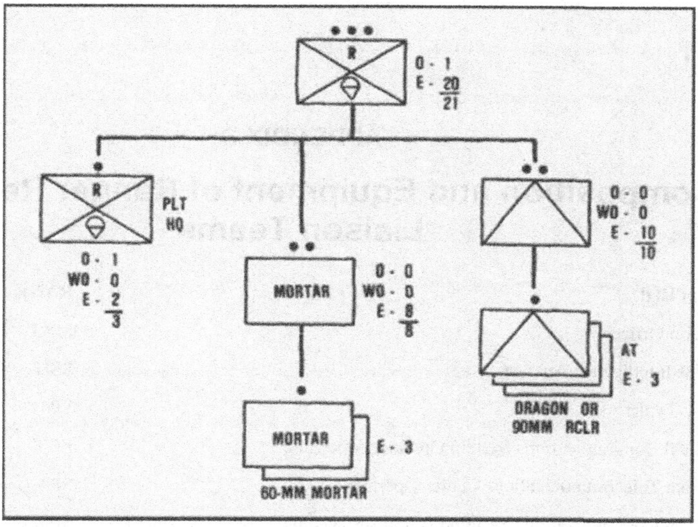

Figure A-11. Ranger weapons platoon.

APPENDIX B

Composition and Equipment of Ranger Regiment Liaison Teams

Team Red	RANK
Liaison Officer	MAJ
Senior Intelligence Analyst	SSG
Clerk-Typist	SP4
Tactical Communications System Operator/Mechanic	SGT
Combat Telecommunications Center Operator	SP4
Tactical Communications System Operator/Mechanic	SP4
Team White	
Liaison Officer	CPT
Senior Intelligence Analyst	SSG
Clerk-Typist	SP4
Tactical Communications System Operator/Mechanic	SGT
Combat Telecommunications Center Operator	SP4
Tactical Communications System Operator/Mechanic	SP4

Equipment (for Each Team)

HMMWV with 3/4-ton trailer	5-kw generator
Communications shelter	Tent, GP, small
AN/PCS-3 radio set (3)	AN/VSC-7 net control station (1)
MX 300 (4)	KY 57 with battery case (5)
HYX 57 with battery case (4)	HYP 57 (5)

KYK 13 (2)

KY 65 with battery case (1)

KG 84 (1)

H189 (2)

GRC 193 (1)

VGC-74 (1)

PP 1648 (1)

SB-22 switchboard (1)

KYI 18 (1)

KY 65 WLA (1)

Digital message device (1)

AN/PRC-77 radio (1)

Facsimile transmission device (1)

TA 312 with WD1 (5)

APPENDIX C
Equipment Density

C-1. Communication equipment.

	Regimental Headquarters	Battalions (x3)
AN/VSC-7 net control station	7	2
AN/PSC-3 radio set	23	13
AN/PRC-41 radio set	2	5
AN/PRC-104 radio set	9	26
AN/GRC-193A radio set	4	0
AN/PRC-70 radio set	6	0
AN/PRC-77 radio set	22	50
AN/VRC-46 radio set	2	0
AN/VRC-49 radio set	0	2
AN/PRC-68 radio set	0	66
TSEC/KY-57 speech security device	60	68
TSEC/KY-58 speech security device	6	0
TSEC/KY-65 speech security device	11	0
AO-8990/P digital message device group	15	23
AN/GXC-7A tactical FAX device	7	2
AN/UGC-74A(V)3 communications terminal	7	0
SB-22 switchboard	2	0
SB-3614(V)/TT switchboard	2	1

TSEC/KG-84	5	0
SB-993/GT switchboard	0	12

C-2. Weapons and equipment each battalion.

Infrared aiming light	96	Interrogator set, Stinger missile	3
Rifle, M16	451	Grenade launcher, M203	67
Rocket launcher, M202	13	Machine gun, M60	27
Mortar, 60-mm M224	6	Rifle, sniper, M21	24
Recoilless rifle, 90-mm	18	Viewer, infrared	9
Tracker, Dragon	9	Night vision devices (all types)	322
Truck, 1/4-ton	2	Laser target designators	11
Radio (all types)	164	Tactical satellite radios (all types)	15
Squad automatic weapon	54		
Laser range finder	14		

APPENDIX D

Standing Orders, Rogers' Rangers

1. Don't forget nothing.
2. Have your musket clean as a whistle, hatchet scoured, sixty rounds powder and ball, and be ready to march at a minute's warning.
3. When you're on the march, act the way you would if you was sneaking up on a deer. See the enemy first.
4. Tell the truth about what you see and what you do. There is an army depending on us for correct information. You can lie all you please when you tell other folks about the Rangers, but don't never lie to a Ranger or officer.
5. Don't never take a chance you don't have to.
6. When we're on the march we march single file, far enough apart so one shot can't go through two men.
7. If we strike swamps, or soft ground, we spread out abreast, so it's hard to track us.
8. When we march, we keep moving till dark, so as to give the enemy the least possible chance at us.
9. When we camp, half the party stays awake while the other half sleeps.
10. If we take prisoners, we keep 'em separate till we have had time to examine them, so they can't cook up a story between 'em.
11. Don't ever march home the same way. Take a different route so you won't be ambushed.
12. No matter whether we travel in big parties or little ones, each party has to keep a scout 20 yards ahead, 20 yards on each flank, and 20 yards in the rear so the main body can't be surprised and wiped out.
13. Every night you'll be told where to meet if surrounded by a superior force.
14. Don't sit down to eat without posting sentries.
15. Don't sleep beyond dawn. Dawn's when the French and Indians attack.
16. Don't cross a river by a regular ford.

17. If somebody's trailing you, make a circle, come back onto your own tracks, and ambush the folks that aim to ambush you.

18. Don't stand up when the enemy's coming against you. Kneel down, lie down, hide behind a tree.

19. Let the enemy come till he's almost close enough to touch. Then let him have it and jump out and finish him up with your hatchet.

MAJ Robert Rogers

APPENDIX E

The Ranger Creed

R Recognizing that I volunteered as a Ranger, fully knowing the hazards of my chosen profession, I will always endeavor to uphold the prestige, honor, and high "esprit de corps" of the Ranger Regiment.

A Acknowledging the fact that a Ranger is a more elite soldier who arrives at the cutting edge of a battle by land, sea, or air, I accept the fact that as a Ranger, my country expects me to move farther, faster, and fight harder than any other soldier.

N Never shall I fail my comrades. I will always keep myself mentally alert, physically strong, and morally straight and I will shoulder more than my share of the task, whatever it may be. One hundred percent and then some.

G Gallantly will I show the world that I am a specially selected and well-trained soldier. My courtesy to superiors, my neatness of dress, and my care of equipment shall set the example for others to follow.

E Energetically will I meet the enemies of my country. I shall defeat them on the field of battle, for I am better trained and will fight with all my might. Surrender is not a Ranger word. I will never leave a fallen comrade to fall into the hands of the enemy and under no circumstances will I ever embarrass my country.

R Readily will I display the intestinal fortitude required to fight on to the Ranger objective and complete the mission, though I be the lone survivor.

RANGERS LEAD THE WAY

APPENDIX G

Training

G-1. Application and guidance.

a. To prepare for combat, ranger units train on missions derived from their analysis of operational and contingency plans and the ranger mission training plans (MTPs). Training guidance is in doctrinal manuals, field circulars, and soldier's training publications. Training must ensure that the ranger unit is skilled in all basic light infantry skills and in ranger-unique skills. In the final analysis, good training depends upon the leadership and teamwork of ranger unit officers and NCOs.

b. Everything rangers do should prepare them to perform well in combat. Ranger units train IAW the battalion training management system (BTMS) concentrating on basic skills and on live-fire, hands-on training.

c. An effective training program breeds high morale and few discipline-related problems. Some qualities of ranger units that have good training programs are as follows:

(1) Leaders who can solve problems.

(2) A thorough and continuous analysis of the unit's mission.

(3) Validated training objectives.

(4) Attention to the basics (move, shoot, and communicate).

(5) Multiechelon sustainment training.

(6) Standardized procedures.

(7) Physical and mental strength.

(8) Live fire as an integral part of training.

(9) Safety as a basis to all operations.

(10) The use of night and adverse weather and difficult terrain as combat force multipliers,

d. When all training is planned with realism, rangers will be in training situations that require the same attention as in battle. Realism and stress in training result in faster learning and better retention of skills. Training objectives must be clear with measurable standards to determine proficiency. Rangers and ranger units must train to the standards that will be expected in combat.

e. Examples of unit training goals include:

(1) Developing each member of the chain of command to perform his duties with competence, confidence, and pride.

(2) Maintaining small unit integrity during all training and support activities. Accomplishing tasks as units.

(3) Ensuring that all tasks for training are keyed to mission needs and weaknesses. Assets--ranger's time, money, and materiel--are expended only after thorough planning.

(4) Training to standard.

(5) Developing and maintaining discipline and esprit to ensure the abilities of each ranger and all units are achieved.

(6) Emphasizing use of the ranger mission training plan (MTP) for each echelon of the organization.

(7) Emphasizing the basics.

(8) Maintaining a high level of physical fitness.

(9) Working together within the ranger regiment and other elements of the special operations forces community.

(10) Developing a unit atmosphere of quick reaction under stress. Encouraging and rewarding individual initiative. Accomplishing unit missions IAW the commander's intent.

G-2. Philosophy.

a. To be effective, training should be physically and mentally challenging. Each ranger will then have a sense of purpose, accomplishment, and satisfaction. Training is evaluated by the results rather than hours of training. Units that are well trained today need reinforcement training to sustain readiness. Training management must be continuous. Training managers must use the time available to improve training efficiency. Cyclic training will not produce the desired results. The "train-to-correct deficiencies" method is needed. The training managers must analyze weaknesses, conduct prescriptive and reinforcement training, and reevaluate their unit's training status. The process that is used to develop and maintain a continuous training and evaluation program is shown in Figure G-1.

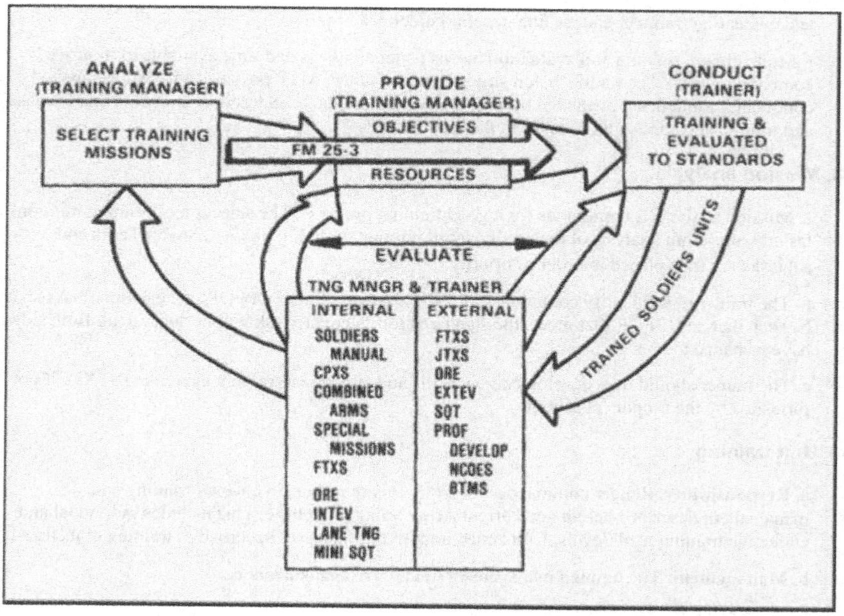

Figure G-1. The Army training and evaluation program.

b. Individual and collective training must be integrated within a unit training program. Training of the individual ranger, leaders, and units at each level must occur at the same time. The ranger commander or leader plans and conducts training for every member of his unit.

c. Ranger commanders analyze operational and contingency plans to determine unit missions and translate the associated tasks into training objectives. These objectives form the basis of the unit training program. The regimental headquarters provides training guidance in a formal regimental training plan. Commanders at lower levels program, supervise, and conduct training within their units.

d. All members of the chain of command are responsible for training. The NCO leader or first-line supervisor conducts individual and small unit collective task training. The officer plans individual training in his unit and directs and conducts collective training.

e. Centralized development of training objectives and decentralized training builds well-trained units and outstanding leaders. Authority and responsibility for the detailed planning, scheduling, conducting, evaluating, and supervising of training are delegated to the lowest element capable of managing this training. Higher headquarters is responsible for establishing objectives, monitoring and evaluating training, and issuing general guidelines.

f. Multiechelon training and evaluation means using all assets and time available to improve combat readiness. The multiechelon structure of the ranger MTP provides a basis for units to be continually trained and evaluated by internal evaluation. Each leader must know his mission tasks and train to accomplish those tasks to the set standards.

G-3. Mission analysis.

a. Mission analysis is continuous for a good training program. The selection of training missions or tasks results from analysis of the unit's overall mission. (See Figure G-1 above.) Tasks and subtasks are then placed in order of priority.

b. The trainer must identify countertasks for the opposing force (OPFOR). The countertasks should be such that an OPFOR that meets the standards for its countertask will be successful. Both sides have a chance to win.

c. The trainer should then develop combat drills and situational training exercises (STXs), if not furnished by the proponent school.

G-4. Unit training.

a. **Responsibility.** Ranger commanders at all levels are responsible for all training and management described herein and current Army training doctrine. This includes individual and collective training at all levels. Units must train to the standards specified in training objectives.

b. **Management.** The training management model for commanders is:

- Set objectives.
- Provide resources.
- Coach subordinates.
- Measure results.
- Provide feedback.

G-5. Individual training.

There are three elements basic to a well-trained individual ranger. He must be physically fit, be proficient in the use of his weapon, and have the psychological commitment to fight the enemy.

a. Individual training is the responsibility of the first-line supervisor under the supervision of the chain of command. It is conducted using performance-oriented techniques under battlefield conditions. New skills must be developed and practiced, and acquired skills must be reinforced throughout the training year. A high level of proficiency must be maintained through continuous training and practice.

b. Trainers should use peer instruction. Commanders ensure needed resources are available to conduct challenging individual training. Ranger units schedule and conduct this training and evaluate individual proficiency.

c. Flexibility in planning and scheduling is necessary. The goal of all individual training must provide for reinforcement training to offset the effects of the forgetting curve. (See Figure G-2).

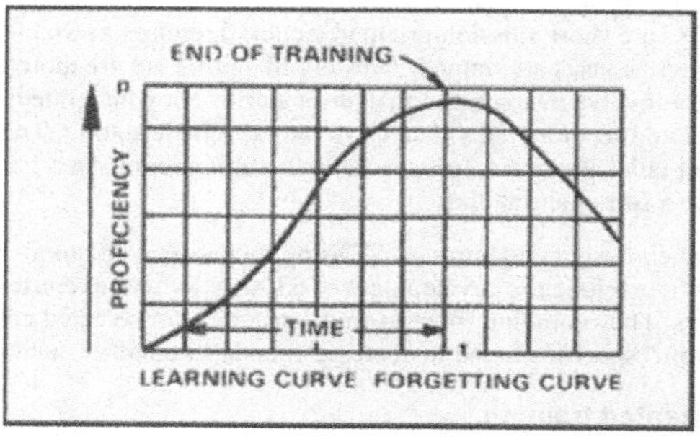

Figure G-2. The learning curve and forgetting curve.

d. The command sergeants major, unit first sergeants, and staff section sergeants are the keys to an effective training program. These senior NCOs must counsel, guide, and critique junior NCOs in training preparation and execution. Fire team leaders, section chiefs, and squad leaders coordinate and present individual training.

e. The NCO supervisors must ensure that the ranger is trained in the skills required for the specific military occupational specialty (MOS) and in those skills used daily. Not all training can be acquired in an office. For example, all rangers must maintain a high skill level with their weapons, be trained in NBC subjects, and be capable of performing contingency missions.

G-6. Collective training.

a. **Guidance.** Collective training trains crews, teams, squads, platoons, companies, and battalions to do their missions. The ranger MTP is used to plan training and evaluate the unit's weaknesses and strengths. If a ranger commander receives new training requirements or missions, they should become part of the collective training program. The ranger MTP describes the combat tasks to be performed by every echelon of the ranger battalion. It is a plan for organizing, conducting, and evaluating the year-round unit training program.

b. **Responsibility.** The ranger unit commander is responsible for the conduct of collective training and for all other training in his unit. Practical exercises emphasize the techniques of terrain use, movement, and effective employment of weapon systems. Training under stress will be conducted to move the unit to a higher plane of combat readiness.

c. **Drills.** Drills are written for small units--for example, fire team and squad, crew and section--and describe an action that is shorter than an entire STX. Drill training objectives cover both individual and collective skills (critical collective tasks) and list them together for effective group action. Drills must be mastered in order to react to rapidly changing situations in combat. The trainer can interrupt the execution of a drill to provide necessary training on individual or collective skills.

d. **Situational training exercise.**

(1) The STXs are short, mission-oriented tactical exercises in which a group of closely related collective tasks are trained. They are like drills but are more complex and flexible. They usually involve a larger unit than does a drill. They integrate leader training, leader tasks, drills, and separate individual tasks into a realistic setting. Trainers may execute STXs as written or tailor them to meet specific unit requirements. An STX is not standardized, but it must have a firm doctrinal base.

(2) Due to their design and purpose, STXs apply mainly to platoon- and company-size units. This does not preclude the development of STXs or similar exercises for unit staffs and headquarters. They combine "how-to-train" specifics for selected collective tasks with the "what-to-train" specifics found in doctrinal manuals and other documents.

G-7. Performance-oriented training.

a. Performance-oriented training emphasizes hands-on performance. Training is conducted to ensure a level of proficiency, rather than established number of hours of required training.

b. Performance-oriented training has precise training objectives and makes good use of resources. Training objectives serve as the basis for preparing, conducting, and evaluating all training.

G-8. Evaluation.

a. A training program must include an evaluation of training (planned and conducted). An evaluation plan must:

(1) Provide immediate feedback to the trainer.

(2) Recognize effective trainers.

(3) Evaluate all units in the conduct of training to provide for a comparison of training programs.

(4) Be flexible to unit needs; provide general and specific evaluations based on the level of training within the unit.

(5) Be in effect at all times.

b. Major considerations for training evaluations include:

(1) Ranger unit commanders establish and maintain an effective internal and external evaluation program.

(2) The director of each staff element is responsible for training evaluations within his area of staff responsibility with emphasis on low-density MOS groups.

(3) The operations officer of the ranger organization develops and validates training and evaluation plans.

c. Evaluation of training must be done vertically from the major task (mission) down through subtasks (collective tasks) to the leader and individual tasks. In selection of individual tasks for evaluation, priority should be given to critical individual tasks that impact on the success of the collective task or mission.

d. The critique and after-action review, after each major training event, are the most important aspects of training. The after-action review takes place immediately after completion of a training task (individual or collective). The individual(s) self-analyzes and critiques the training event. Then after a formal critique of the training, the individuals) trains to standard (not to time).

G-9. Field training.

a. **Considerations.** Ranger commanders schedule training under field conditions to permit maximum training opportunities. Field training emphasis is on mastery of a series of individual (soldier and leader) tasks and collective training tasks. These tasks are repeated until established standards are met. The goal of ranger field training is to ensure realistic, demanding, and meaningful training.

(1) When training is conducted under conditions of moderate stress, most combat critical skills are learned quicker and retained longer. (See Figure G-3 on the following page.) Stress can be incorporated in a training situation by physical deprivation (sleep, food), by live-fire exercises, by operation against an OPFOR under simulated battlefield conditions for an extended period, by the intensity of activity including demands for decision making, or by physically demanding activities. Ranger units train under stress so that they will react quickly in combat.

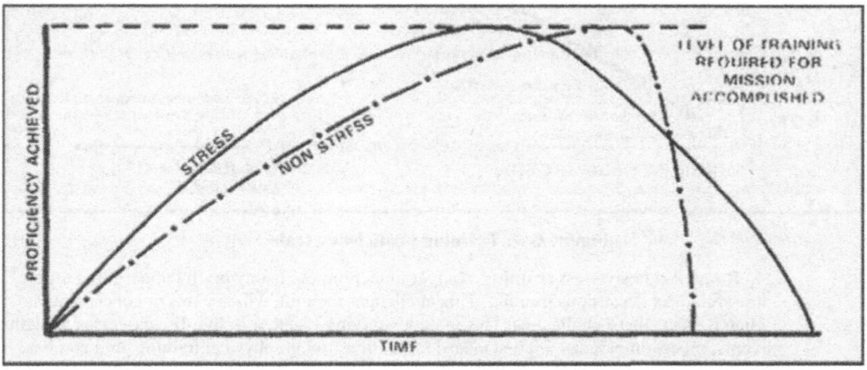

Figure G-3. The Army training and evaluation program.

(2) Antagonistic training situations are two-sided exercises conducted by opposing elements. Each element has a training objective (task and standards). The opposing force has a corresponding countertask. This training may be progressive--for example, starting with a

crawl-through, then a walk-through, and progressing to both elements operating at a fast pace. The use of multiple integrated laser equipment system (MILES) greatly enhances task and countertask training. The units should also be considered for a live-fire exercise against a realistic OPFOR target array when the basics have been mastered.

b. **Methods of training.**

(1) **Considerations.** The chance for success in combat can be increased by becoming proficient in training situations that depict realistic combat conditions. (See Figure G-4.) Meeting the standards of a realistic live-fire training exercise assures the commander that his unit could perform well in combat. Figure G-4 shows the relationship between realistic training and the challenge to the trainer.

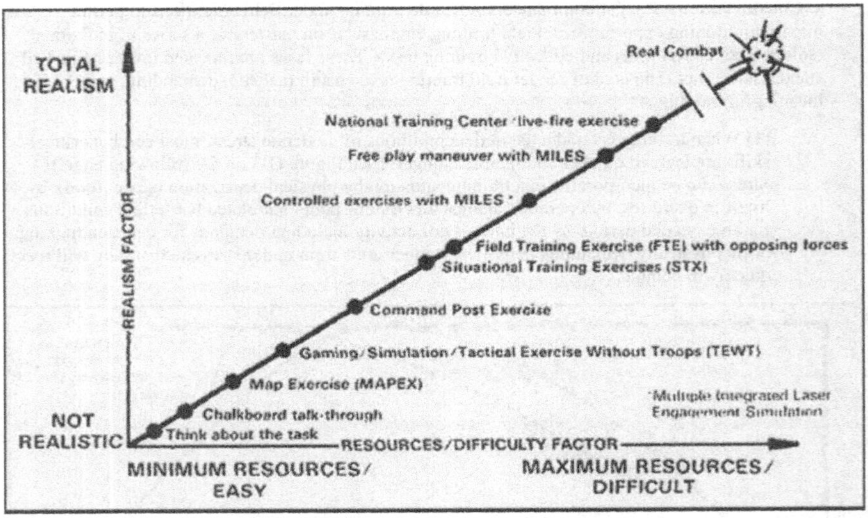

Figure G-4. Training proficiency scale.

(2) **Ranger cohesiveness training.** Ranger units promote teamwork by undergoing tough, stressful, often hazardous training. This challenges the unit, which gains in cohesiveness when it meets these challenges. This includes serving together in live-fire exercises, athletic events, precision parades, high-standard inspections, tough physical training, and combat.

(3) **Field training exercise.** An FTX is training normally conducted under simulated, realistic combat conditions. This training normally exercises command and control systems and the administrative and logistical needs of sustained combat. Stress training is part of this exercise, focusing on live-fire and MILES drills and STXS.

(4) **Combined arms training.** Combined arms training should include all means of support such as tactical air, armed helicopters, and indirect fire with emphasis on fire support coordination and airspace control.

(5) **Live-fire exercises.** Including live-fire exercises in drills and STXs offers a chance to evaluate units in their response to simulated combat missions. Ranger units train under pressure and always anticipate being selected to conduct a live-fire exercise. Live-fire exercises develop confidence and esprit, and reinforce unit discipline.

(6) **Opposing force.** Committing one unit against another adds a competitive spirit to a field training exercise (FTX). The use of MILES and squad combat operations exercises (simulation) (SCOPES) provides realism to two-sided field exercises. Trainers should use OPFOR equipment and training devices available, as well as making their own devices. (See FM 30-103, Appendix R, for more information.)

c. **Key functional training areas.**

(1) **Land navigation and patrolling.** Day and night navigation and patrolling exercises must be emphasized. Orienteering improves these skills, as well as physical fitness.

(2) **Movement techniques.** Infantry combat has two major components--to move and to hit the enemy first. Ranger units must be skilled in the use of traveling, traveling overmatch, and bounding overmatch. Units in contact must be skilled in moving under fire.

(3) **Fighting** positions. Training includes the building of concealed fighting positions.

(4) **Mine and countermine training.** Training includes how to defend against hostile mines, booby traps, and other demolitions. The identification, installation, neutralization, and removal of US mines and devices is also stressed.

(5) **Limited visibility operations.** Ranger units must be skilled in limited visibility operations and the use of night vision and thermal devices.

(6) **Operation security.** Operation security includes deception, physical security, signal security (SIGSEC), and information security. It is taught in all training and at all levels. Although all areas of OPSEC must be measured in training, techniques of camouflage and SIGSEC need special emphasis. (See AR 530-1 and TC 100-1 for more information.)

G-10. Professional development.

Ranger units at battalion and regimental levels conduct many officer and NCO professional development programs. This ensures that all officers and NCOs are skilled in their field, confident, and wellrounded. Professional readings and book reports, tactics seminars, and tactical exercises without troops are good development methods.

G-11. Physical training.

a. This program ensures that each ranger maintains the physical ability and stamina needed to do his mission. Other objectives are to promote unit cohesiveness and discipline. These are met by combining individual testing, formation, running, athletic competition, weight control, proper diet, teamwork drills, and freedom from drug and alcohol abuse. Mental fitness and toughness come from experience and success. Good physical conditioning increases a unit's chances of success.

b. Physical training (PT) programs are designed to include the elements of overload, progression, balance, variety, and regularity.

c. Ranger unit physical fitness standards are:

(1) Score 80 points for each Army physical readiness test (APRT) event and do six chinups.

(2) Pass the ranger swimming test.

(3) Complete an 8-kilometer run in 40 minutes.

(4) Complete a 12-kilometer road march in three hours (with rucksack, helmet, and weapon).

(5) Meet the Army height and weight standards.

G-12. Environmental and specialized training.

a. Rangers conduct training in environments as close to real combat missions as possible. This conditions rangers to survive and fight in a specific environment. Such training includes combat-critical tasks and technique training for that environment.

b. Each ranger battalion's environmental and specialized training includes at least the following:

(1) Jungle (once every 12 months).

(2) Desert (once every 12 months).

(3) Urban (once every 6 months).

(4) Extreme cold (once every 18 months).

(5) Mountain (once every 12 months).

(6) Amphibious (once every 18 months).

G-13. Ranger indoctrination program.

a. The RIP is a training and selection program. It is run by specially selected NCOs from the regimental headquarters. It has as its mission the training and indoctrination of newly assigned soldiers. It physically toughens and mentally sharpens the newly assigned soldier. It teaches basic skills and techniques. The RIP identifies and eliminates in three weeks any applicant who does not show dedication, motivation, physical fitness, and emotional stability.

b. An indoctrination program for newly assigned ranger-qualified officers and NCOs prepares them to assume duty as leaders within the ranger regiment. These personnel monitor and participate in RIP training, as well as undergoing individual ranger leadership training. The RIP provides them an overview of the regiment's history, its mission, and its goals. (See Appendixes C,

D, and E for further information about the history and goals of the ranger regiment.) The RIP also provides a review of the skills required in a ranger unit. Refresher training in airborne and pathfinder operations, use of indirect fires, and light infantry doctrine is conducted. During RIP all newly assigned officers and NCOs must meet the ranger physical fitness standard.

c. The RIP cadre also conducts a preranger program to prepare soldiers to attend the US Army Ranger School. It is a physically demanding course that teaches troop-leading procedures, operation orders, and small unit patrol actions.

Glossary

The following acronyms, abbreviations, and definitions are provided to aid the reader in understanding this publication.

Section I

ACRONYMS AND ABBREVIATIONS

AAT	Army assault team
ABCCC	airborne battlefield command and control center
ACL	allowable cargo load
ADA	air defense artillery
AFFOR	the USAF element of a joint task force
AGL	above ground level
AKZ	armor kill zone
ALCC	airlift control center
ALCE	airlift control element
AM	amplitude modulation
AMC	air mission commander
APERS	antipersonnel
APRT	Army physical readiness test
ARFOR	the Army element of a joint task force
ARSOC	Army special operations command
ARSOF	Army Special Operations Forces
ASG	area support group
ASPC	all-source production center

ATGM	antitank guided missile
AWADS	adverse weather aerial delivery system
BCE	battle coordination element
BTMS	battalion training management system
CA	civil affairs
CAS	close air support
CCT	combat control team
CDS	container delivery system
CE	communications-electronics
CI	counterintelligence
CINC	commander in chief
COMINT	communications intelligence
COMSEC	communications security
CONUS	continental United States
COSCOM	corps support command
CP	command post
CS	combat support; a chemical agent
CSS	combat service support
DACG	departure airfield control group
DACO	departure airfield control officer
DASC	direct air support center
DMDG	digital message device group
DOD	Department of Defense
DS	direct support
DSU	direct support unit
DZ	drop zone
EAC	echelons above corps

EACIC	Echelons Above Corps Intelligence Center
ECCM	electronic counter-countermeasure
ECM	electronic countermeasures
E&E	escape and evasion
EEFI	essential elements of friendly information
EOC	emergency operations center
EW	electronic warfare
FARP	forward area rearm and refuel point
FASCAM	family of scatterable mines
FAX	facsimile transmission
FDC	fire direction center
FEBA	forward edge of the battle area
FFL	friendly front lines
FIST	fire support team
FLOT	forward line of own troops
FM	field manual; frequency modulation
FO	forward observer
FORSCOM	US Army Forces Command
FPF	final protective fire
FPL	final protective line
FRAGO	fragmentary order
FSA	fire support area
FSCOORD	fire support coordination or coordinator
FSE	fire support element
FSO	fire support officer
FSS	fire support station
FTX	field training exercise

G3	Assistant Chief of Staff, Operations
GS	general support
GSR	ground surveillance radar
HAARS	high-altitude, airdrop resupply system
HAHO	high-altitude, high-opening
HALO	high-altitude, low-opening
HE	high explosive
HF	high frequency
HHC	headquarters and headquarters company
HSLLADS	high-speed, low-level airdrop system
HUMINT	human intelligence
IAW	in accordance with
IFF	identification-friend-or-foe
IP	initial point
IRP	initial rally point
ISB	intermediate staging base
JACC/CP	joint airborne communications center/command post
JCS	Joint Chiefs of Staff
JCSE	joint communications support element
JFC	joint force commander
JSOC	joint special operations command
JTB	joint transportation board
JTF	joint task force
kw	kilowatt
LAPES	low altitude parachute extraction system
LAW	light antitank weapon
LNO	liaison officer

LOC	logistical operations center
LRRP	long-range reconnaissance patrol
LRSU	long-range surveillance unit
LTD	laser target designator
LZ	landing zone
MAC	Military Airlift Command
MANPADS	man-portable air defense system
MCC	movement control center
METT-T	mission, enemy, terrain, troops and time available
MI	military intelligence
MILES	multiple integrated laser engagement system
mm	millimeter
MMC	Materiel Management Center
MMEE	minimum mission-essential equipment
MOS	military occupational specialty
MOUT	military operations on urbanized terrain
MRLS	multiple rocket launch system
MTP	mission training plan
NBC	nuclear, biological, chemical
NCA	National Command Authority
NCO	noncommissioned officer
NOD	night observation device
OCOKA	observation, concealment, obstacles, key terrain, and avenues of approach
OCONUS	outside continental United States
OPCOM	operational command
OPCON	operational control
OPFOR	opposing force

OPORD	operation order
OPSEC	operations security
ORP	objective rally point
PAC	personnel administration center
PDF	primary direction of fire
PIR	priority intelligence requirements
PLL	prescribed load list
POM	preparation for oversea movement
PSYOP	psychological operations
PT	physical training
PZ	pickup zone
QRE	quick-reaction element
RATELO	radiotelephone operator
REMAB	remote marshalling base
RFL	restrictive fire line
RIP	ranger indoctrination program
RP	rally point; release point
RPG	rocket-propelled grenade
RRF	ranger ready force
RRP	reentry rally point
RSE	ranger support element
RSTA	reconnaissance, surveillance, and target acquisition
S1	adjutant
S2	intelligence officer
S3	operations and training officer
S4	supply officer
S5	civil affairs officer

SAR	search and rescue
SATCOM	satellite communications)
SCIF	sensitive compartmented information facility
SCOPES	squad combat operations' exercises (simulation)
SCUBA	self-contained underwater breathing apparatus
SEAD	suppression of enemy air defense
SEMA	special electronic mission aircraft
SERE	survival, evasion, resistance, and escape
SF	special forces
SHORAD	short-range air defense
SIGSEC	signal security
SKE	station-keeping equipment
SLAR	side-looking airborne radar
SOA	special operations aviation
SOCB	special operations communications battalion
SOCOM	special operations command (usually, 1st SOCOM)
SOP	standing operating procedures
SOSB	special operations support battalion
SOTF	special operations task force
SPL	support platoon leader
STX	situational training exercise
TA	theater Army
TAACOM	Theater Army Area Command
TAC	Tactical Air Command
TACP	tactical air control party
TACSAT	tactical satellite terminal
TAMMC	Theater Army Materiel Management Center

TASG	theater Army support group
TDA	table of distribution and allowances
TOA	transfer of authority
TOC	tactical operations center
TOE	table(s) of organization and equipment
TRP	target reference point
TTY	teletype
UHF	ultrahigh frequency
US	United States
USAF	United States Air Force
USMC	United States Marine Corps
USN	United States Navy
VHF	very high frequency
WP	white phosphorus
XO	executive officer

Section II

DEFINITIONS

ADVANCE FORCE. An operation requiring the immediate deployment of US forces with the intent to negate a threat before the deployment of follow-on forces is required.

AIRLAND BATTLE DOCTRINE. An approach to military operations that realizes the full potential of US forces. Two notions (extending the battlefield and the ability to integrate conventional, nuclear, chemical, and electronics means) are combined to describe a battlefield where the enemy is attacked to the full depth of his formation.

AREA OF INFLUENCE. A geographical area wherein a commander is directly capable of influencing operations by maneuver or fire support systems normally under his command or control.

AREA OF INTEREST. That area of concern to the commander, including the area of influence, areas adjacent thereto, and extending into enemy territory to the objectives of current or planned operations. This area also includes areas occupied by enemy forces who would jeopardize the accomplishment of the mission.

AREA OF OPERATIONS. That portion of an area of conflict necessary for military operations. Areas of operations are geographical areas assigned to commanders for which they have responsibility and in which they have authority to conduct military operations.

CHALK. One specific aircraft load.

CLANDESTINE OPERATIONS. Activities to accomplish intelligence, counterintelligence, and other similar activities sponsored or conducted by governmental departments or agencies, in such a way as to ensure secrecy or concealment. (They differ from covert operations in that emphasis is placed on concealment of the operation rather than on concealment of the identity of the sponsor.)

COMMUNICATIONS INTELLIGENCE (COMINT). The interception and processing of foreign communications passed by radio, wire, or other electromagnetic means, and by the processing of foreign encrypted communications, however transmitted. Interception comprises search, intercept, operator identification, signal analysis, traffic analysis, crypto-analysis, study of plain text, the fusion of these processes, and reporting the results.

COMMUNICATIONS SECURITY. The protection resulting from all measures designed to deny to unauthorized persons information of value that might be derived from the possession and study of telecommunications, or to mislead unauthorized persons in their interpretations of the results of such study. COMSEC includes crypto-security, physical security, and transmission security.

COUNTERINTELLIGENCE. That aspect of intelligence activities, both offensive and defensive, designed to detect and neutralize or destroy the effectiveness of hostile foreign intelligence activities and to protect information against espionage, personnel against subversion, and installations and materiel against sabotage.

COVERT OPERATIONS. Operations that are so planned and executed as to conceal the identity of or permit plausible denial by the sponsor. They differ from clandestine operations in that emphasis is placed on concealment of identity of the sponsor rather than on concealment of the operation.

DECEPTION. Those measures designed to mislead the enemy by manipulation, distortion, or falsification of evidence to induce him to react in a manner prejudicial to his interests.

DEMONSTRATION. A show of force in an area where a decision is not sought. It is similar to a feint but does not make contact with the enemy.

DRILL. A drill focuses on a "chunk" or "slice" of battle and is targeted at small units. Drills are critical collective tasks that require a high degree of proficiency from a small unit. Generally, small units "drill" collective tasks that require rapid responses by unit members in the absence of detailed orders from unit leaders.

ELECTRONIC COUNTERMEASURES. That division of electronic warfare involving actions taken to prevent or reduce an enemy's effective use of the electromagnetic spectrum.

ELECTRONIC INTELLIGENCE (ELINT). The technical and intelligence information derived from foreign noncommunications electromagnetic radiations emanating from other than atomic detonations or radioactive sources.

ELECTRONIC WARFARE. Military action involving the use of electromagnetic energy to determine, exploit, reduce, or prevent hostile use of the electromagnetic spectrum and action that retains friendly use of the electromagnetic spectrum.

ELECTRONIC WARFARE SUPPORT MEASURES (ESM). That division of EW involving actions taken to search for, intercept, identify, and locate sources of radiated electromagnetic energy for the purpose of immediate threat recognition. ESM provides a source of information needed for immediate action involving electronic countermeasures, electronic counter-countermeasures, avoidance, targeting, and other tactical employment of forces.

ESPIONAGE. Actions directed toward the acquisition of information through clandestine operations.

EVACUATION/SAFEGUARD. The use of US forces to evacuate US citizens or allies from a hostile or potentially hostile area, or to safeguard US personnel or property.

EXFILTRATION. The removal of personnel or units from areas under enemy control by stealth, deception, surprise, or clandestine means. Exfiltration is a method of extracting ranger units.

EXTRACTION. The removal of personnel or units from areas under enemy control by any means, to include exfiltration, airland, airmobile, or amphibious.

FEINT. A diversionary supporting attack conducted to draw the enemy's attention from the main effort. It is normally executed by brigades and small units. Feints are usually shallow, limited objective attacks that go in before or during the main attack.

GUERRILLA WARFARE. Military and paramilitary operations conducted in enemy-held or hostile territory by irregular and predominantly indigenous forces.

HIGH-INTENSITY CONFLICT. A war between the US and any other nation or nations in which both belligerents use the most modern technology and all resources in intelligence, mobility, firepower (including NBC), command, communications and control, and service support.

HUMAN INTELLIGENCE. Intelligence derived from information provided by human resources.

INFILTRATION. The movement through or into an area of territory occupied by either friendly or enemy troops or organizations. The movement is made either by small groups or by individuals, at extended or irregular intervals. When used in conjunction with the enemy, it implies that contact is avoided. Infiltration is a subset of insertion.

INSERTION. Placement of troops and equipment into an operational area by any means.

INTERDICT. To isolate or seal off an area by military means; to prevent, hinder, or delay the use of an area or route by enemy forces.

INTERDICT LINE OF COMMUNICATIONS. An attack to seal off an area or to deny use of a route or approach.

JOINT TASK FORCE. A JTF may be constituted and designated by the Secretary of Defense or by the commander of a unified command, specified command, or an existing JTF. Normally, it performs missions having specific, limited objectives or missions of short duration. It dissolves when it has achieved its purpose. The joint task force commander is responsible to the commander, known as the establishing authority, who created the JTF. The JTF is composed of elements of two or more services operating under a single JTF commander. The JTF commander has operational control over the entire force and may have direct command of his own staff with representatives from the other services. He exercises logistics coordination or control only as necessary to meet his subordinate commanders' logistics needs.

LIAISON. That personal contact or communication maintained between elements of military forces to ensure mutual understanding and unity of purpose and effort.

LINE OF COMMUNICATIONS. All routes (land, water, and air) that connect an operating military force with one or more bases of operations and along which supplies and military forces move.

LOW-INTENSITY CONFLICT. Limited politico-military operations conducted to achieve political, social, economic, or psychological objectives. It is generally confined to a specific geographic area and is often characterized by operational constraints imposed on ranger units by political and economic considerations. It may involve intermittent combat against regular threat military forces, but it more often involves conflict with irregular, security, or paramilitary forces of a country or group involved in an operation that is counter to the best interests and national security of the United States. It is characterized by relatively low levels of combat and constraints placed on the weaponry, tactics, and movements of the ranger force. It may involve counterterrorist or antiterrorist operations and generally occurs before a pro forma declaration of war.

MID-INTENSITY CONFLICT. War, declared or undeclared, between the US and the regular forces of one or more nations and their respective allies, if any, in which the belligerents use the most modern technology and all resources in intelligence, mobility, firepower (excluding nuclear, chemical, and biological weapons), command, communications and control, and support. They are used for limited objectives under definitive policy limitations as to the extent of destructive power that can be used or the extent of geographical area that might be involved.

MILITARY STRATEGY. Art and science of using the armed forces of a nation to secure the objectives of national policy by application of force or the threat of force. Military strategy sets the fundamental conditions for operations.

NATIONAL COMMAND AUTHORITY. This includes the President and Secretary of Defense or the duly deputized alternates or successors. Commonly referred to as NCA.

OPERATIONAL COMMAND/CONTROL. Operational command (OPCOM) and operational control in joint force terminology both refer to the authority exercised by joint commanders over subordinate service components. Those terms are not interchangeable in joint operations. OPCOM applies to the authority exercised by commanders of unified commands. OPCON is the authority that subordinate joint task force commanders exercise in the conduct of specific operations. The authority that military departments exercise over their respective components is commonly referred to as command less operational command.

OPERATIONAL LEVEL OF WAR. The operational level of war uses available military resources to attain strategic goals within a theater of war. It is the theory of larger unit operations. It also involves planning and conducting campaigns. Campaigns are sustained operations designed to defeat an enemy force in a specified space and time with simultaneous and sequential battles. The disposition of forces, selection of objectives, and actions taken to weaken or to outmaneuver the enemy all set the terms of the next battle and exploit tactical gains. They are all part of the operational level of war. In AirLand Battle doctrine, this level includes the marshalling of forces and logistical support, providing direction to ground and air maneuver, applying conventional and nuclear fires in depth, and employing unconventional and psychological warfare.

OPERATIONS SECURITY. Those measures designed to protect information concerning planned, ongoing, and completed operations from unauthorized disclosure. It includes all actions a command takes to deny the enemy information about friendly units and their operations.

ORDER OF BATTLE (OB). The identification, strength, command structure, and disposition of the personnel, units, and equipment of any military force.

RAID. Usually a small-scale operation, involving a swift penetration of hostile territory to secure information, confuse the enemy, or destroy his installations. It ends with a planned withdrawal upon completion of the assigned mission.

RANGER MISSION. To conduct strike, raid, or special light infantry operations. Execution of ranger missions is normally directed by the National Command Authority. Rangers also support operational-level commanders during operations of limited duration when allocated or assigned by the theater commander.

RECONNAISSANCE. A mission undertaken to obtain, by visual observation or other detection methods, information about the activities and resources of an enemy or potential enemy; or to secure data concerning the meteorological, hydrographical, or geographical characteristics of a particular area.

SEARCH AND RESCUE. The use of aircraft, surface craft, submarines, specialized rescue teams, and equipment to search for and rescue personnel in distress on land or at sea.

SEIZURE. As an operational purpose, the capture of a voluntarily restricted portion of an enemy-controlled territory. Capture of an isolated land mass such as an island is usually categorized as a seizure.

SEIZE KEY FACILITY. An attack on an installation with the intent of holding it for a short period. Normally, rangers would make the assault to seize and secure an installation for follow-on forces.

SERVICE COMPONENT. Each service component commander is responsible for recommending the proper use of his forces and for accomplishing operational tasks assigned by the joint commander. He is also responsible for his service in matters of internal administration and discipline; training in own service doctrine, techniques, and tactics; designation of specific units to meet joint requirements; logistics functions normal to the component; tactical employment of service component forces; and service intelligence matters.

SHOW OF FORCE. A mission carried out to demonstrate US resolve, whereby US forces are deployed to defuse a situation that may be harmful to US interests or national objectives.

SIGNAL INTELLIGENCE (SIGINT). A generic term that includes COMINT and ELINT. Activities concerned with the collection and processing of foreign electromagnetic emissions for the purpose of producing intelligence information.

SITUATIONAL TRAINING EXERCISE. An STX is a short, mission-oriented exercise in which a group of closely-related collective tasks are trained, using a doctrinally preferred method of executing these tasks to established standards. An STX is drill-like in nature, but it is more complex and flexible than a drill. It usually involves a larger unit than a drill.

SPECIAL OPERATIONS. Military operations conducted by specially trained, equipped, and organized DOD forces against strategic or tactical targets in pursuit of national, military, political, economic, or psychological objectives. Special operations may support conventional military operations, or they may occur independently when the use of conventional forces is either inappropriate or infeasible. Sensitive peacetime operations, except for training, are normally authorized by the NCA and conducted under the direction of the NCA or designated commander. Special operations may include unconventional warfare, counter-terrorist operations, collective security, psychological operations, and civil affairs measures.

SPECIAL OPERATIONS FORCES. All civil affairs, psychological operations, ranger (active component), and special forces units within the US Army, US Army Reserve, and National Guard.

SPECIAL OPERATIONS COMMAND. The major subordinate command to FORSCOM exercising command and control of all active component Army special operations forces not placed under the operational command of unified or specified commands.

STRATEGIC LEVEL OF WAR. Military strategy uses the armed forces of a nation to secure the objectives of national policy by applying force or the threat of force. Military strategy sets the fundamental conditions for operations.

STRIKE. An attack that is intended to inflict damage on, seize, or destroy an objective. Sometimes referred to as direct action missions.

SURVEILLANCE. The systematic observation of aerospace, surface or subsurface areas, places, persons, or things by visual, aural, electronic, photographic, or other means.

TACTICAL. Pertaining to the employment of units in combat.

TACTICAL LEVEL OF WAR. Tactics are the specific techniques smaller units use to win battles and engagements that support operational objectives. Tactics employ all available combat, combat support, and combat service support. Tactics involve the movement and positioning of forces on the battlefield in relation to the enemy, the provisioning of fire support, and the logistical support of force before, during, and following engagements with the enemy. At corps and division, operational and tactical levels are not clearly separable, but they are guided by the same principles. An operation designed to defeat any enemy force in an extended area does so through operational maneuver and a series of tactical actions.

TACTICAL COMBAT INTELLIGENCE. That knowledge of the enemy, weather, and geographical features needed by a commander in the planning and conduct of combat operations. It is derived from the interpretation of information on the enemy and the environment.

TARGET ANALYSIS. A detailed and systematic examination of processed intelligence to identify and locate targets.

UNIFIED COMMAND. A unified command is established and designated by direction of the President. The command is organized under a single unified CINC and has a broad, continuing mission. The CINC is responsible to the Secretary of Defense for accomplishing his broad and continuing mission and operates under the strategic and operational direction of the JCS. A unified command is composed of assigned forces of two or more services. These forces, as service components, are under the operational command of the unified commander. The CINC exercises operational command authority through service component commanders. When tasked by higher authority, he may establish a subordinate JTF to conduct specific missions. Under emergency conditions, the CINC has the authority to use all facilities and supplies of assigned forces to accomplish his mission.

References

REQUIRED PUBLICATIONS

Required publications are sources that users must read in order to understand or to comply with this publication.

Field Manuals (FM)

FM 7-8	The Infantry Platoon and Squad (Infantry, Airborne, Air Assault, Ranger)
FM 7-10	The Infantry Rifle Company (Infantry, Airborne, Air Assault, Ranger)
FM 7-20	The Infantry Battalion (Infantry, Airborne, and Air Assault)

RELATED PUBLICATIONS

Related publications are sources of additional information. They are not required in order to understand this publication.

Army Regulations (AR)

AR 350-1	Army Training

Field Manuals (FM)

FM 10-52-1	Commander's Handbook for Water Usage in Desert Operations
FM 10-501	Airdrop of Supplies and Equipment: Rigging Containers
FM 10-512	Airdrop of Supplies and Equipment: Rigging Typical Supply Loads
FM 10-542	Airdrop of Supplies and Equipment: Rigging Loads for Special Operations
FM 10-547	Airdrop of Supplies and Equipment: Rigging the High Speed Aerial Delivery Container CTU-2/A
FM 10-553	Airdrop of Supplies and Equipment: Rigging Ammunition
FM 12-3-4	Echelons Above Corps (EAC) Personnel and Administrative Doctrine

FM 14-6	Comptroller/Finance Services in Theaters of Operations
FM 19-15	Civil Disturbances
FM 20-11-1	Military Diving (Volume 1)
FM 20-11-2	Military Diving (Volume 2)
FM 21-20	Physical Training Program
FM 21-26	Map Reading
FM 21-75	Combat Skills of the Soldier
FM 21-76	Survival, Evasion, and Escape
FM 21-77A(S)	Joint Worldwide Evasion and Escape Manual
FM 22-9	Soldier Performance in Continuous Operations
FM 23-67	Machinegun, 7.62-mm, M60
FM 25-1	Training
FM 25-2	Unit Training Management
FM 25-3	Training in Units
FM 25-4	How to Conduct Training Exercises
FM 25-5	Training for Mobilization and War
FM 25-7	Training Ranges
FM 26-2	Management of Stress in Army Operations
FM 29-47	Main Supply and Service Company, Forward Supply and Service Detachments, Supply and Service Battalion, Airborne Division
FM 29-51	Division Supply and Field Service Operations
FM 30-5	Combat Intelligence
FM 30-17	Counterintelligence Operations
FM 31-18	Long-Range Reconnaissance Ranger Company
FM 31-19	Special Forces Military Free-Fall Parachuting
FM 1-24	Special Forces Air Operations

FM 31-25	Special Forces Waterborne Operations
FM 31-70	Basic Cold Weather Manual
FM 34-1	Intelligence and Electronic Warfare Operations
FM 34-60A	Counterintelligence Operations (U)
FM 55-9	Unit Air Movement Planning
FM 57-38	Pathfinder Operations
FM 57-220	Basic Parachuting Techniques and Training
FM 63-3J	Combat Service Support Operations--Corps
FM 63-4	Combat Service Support Operations--Theater Army Area Command
FM 63-5	Combat Service Support Operations--Theater Army
FM 90-3	Desert Operations
FM 90-4	Airmobile Operations
FM 90-5	Jungle Operations
FM 90-6	Mountain Operations
FM 90-8	Counterguerilla Operations
FM 90-10	Military Operations on Urbanized Terrain (MOUT)
FM 90-10-1	An Infantryman's Guide to Urban Combat
FM 100-5	Operations
FM 100-20	Low Intensity Conflict
FM 100-26	The Air-Ground Operations System
FM 101-5	Staff Organization and Operations
FM 101-10-1	Staff Officers' Field Manual: Organization, Technical and Logistical Data (U)

Firing Table (FT)

FT 60-P-1	Firing Tables for Mortar, 60-mm: M224, Firing Cartridge, HE, M720, Cartridge, HE, M49A4, Cartridge, TP, M5OA3, Cartridge, WP, M302AI, Cartridge, Illuminating, M83A3

Soldier's Training Publication (STP)

STP 7-11B24-SM	Soldier's Manual MOS 11B, Infantryman, Skill Level 2/3/4

Technical Manuals (TM)

TM 10-1670-267-12&P	Operator's and Organizational Maintenance Manual Including Repair Parts and Special Tools List for High Altitude Airdrop Resupply System: 500 Pound Capacity

Training Circulars (TC)

TC 23-14	Sniper Training and Employment
TC 23-90	Mortar Training
TC 24-1	Communications-Electronics Operation Instructions, the CEOI
TC 31-20-5(C)	Special Forces Communications (U)
TC 90-6-1	Military Mountaineering

Other Related Publications

King, Michael J. Dr., *Leavenworth Papers, No. 11*, "Rangers Selected Combat Operations in World War II," Combat Studies Institute, USACGSC, Fort Leavenworth, KS

PROJECTED PUBLICATIONS

Projected publications are sources of additional information that are scheduled for printing but are not yet available. Upon print, they will be distributed automatically via pinpoint distribution. They may not be obtained from the USA AG Publications Center until indexed in DA Pamphlet 310-1.

Field Manuals (FM)

FM 7-70	Light Infantry Squad/Platoon
FM 7-71	Light Infantry Company
FM 7-72	Light Infantry Battalion
FM 7-73	Light Infantry Brigade

FM 23-36	60-mm Mortar, M224

Made in United States
Orlando, FL
04 November 2026

90218329R00197